WATERSHED DEVELOPMENT AND LIVELIHOODS

WATERSHED DEVELOPMENT AND LIVELIHOODS

People's Action in India

S. K. DAS

Routledge
Taylor & Francis Group

LONDON AND NEW YORK

First published 2008 by Routledge

2 Park Square, Milton Park, Abingdon, Oxfordshire OX14 4RN
711 Third Avenue, New York, NY 10017

Routledge is an imprint of the Taylor & Francis Group, an informa business

First issued in paperback 2018

Transferred to Digital Printing 2008

Typeset by
Star Compugraphics Private Limited
5-CSC, 1st Floor,
Vasundhara Enclave,
Delhi 110 096

British Library Cataloguing-in-Publication Data
A catalogue record of this book is available from the British Library

ISBN 978-0-415-44904-5 (hbk)
ISBN 978-1-138-38418-7 (pbk)

At first I was afraid of everyone
and everything: my husband, the
village sarpanch, the police.
Today I fear no one. I have my
own bank account, I am the leader
of my village's saving group…
I tell my sisters about our
movement. And we have a 40,000-
strong union in the district.

From a discussion group of poor men and women
(The World Bank 2000)

Contents

List of Figures

List of Tables

List of Map

Acknowledgements

I acknowledge the generous assistance I received from a large number of people in the course of writing this book. Among those who were closely involved, one name stands out. Without the help of Washim Akhtar, the Collector of Jhabua, this book could not have been written. I am deeply indebted to him for his encouragement, assistance and steady supply of reference material.

Introduction

Jhabua is what Pico Iyer would call a lonely place (Iyer 1993). Tucked away in a far corner of the Vindhya mountains, it has no seat at conference tables. It is as if Jhabua is an exile both in time and space, the landscape is bleak with sandy, brown hills repeating themselves in weary succession.

But for the Bhil who live there it is home. This is where the music of the *dhol* and *bansari* are part of everyday life; where the landscape changes when toddy is ready and the Mahua tree blossoms; where in summer, the Palash flowers turn a violent red it seems as if forest is on fire; and where they still use arrows with expertise. Wasn't Eklavya, the fabled archer in the Mahabharatha, a Bhil?

Eklavya was indeed a Bhil, and legend goes that in his eagerness to learn archery he approached Dronacharya, the greatest teacher in archery, to teach him. But Dronacharya refused because he taught only Kshatriya princes, his rejection did not discourage Eklavya who, in turn, made a clay image of Dronacharya and taught himself archery in front of the image. He went on to become an excellent archer.

However, Eklavya's archery skills did not go unnoticed for long as Dronacharya and his Kshatriya disciples chanced upon him shooting arrows. But it was when he surpassed all the Kshatriya princes at an open competition that Dronacharya realised that Eklavya was even better than Arjuna, his favourite disciple. As he didn't want Arjuna to be overshadowed, he asked Eklavya if he was prepared to give *gurudakshina* (teacher's fee) since he had learnt his skills in front of his clay image. Eklavya agreed, and Dronacharya asked him to cut off his right thumb and give it as *gurudakshina*. He did so, and without his right thumb there was no chance of Eklavya surpassing Arjuna in archery.

As Eklavya was marginalised, so were the Bhil through the centuries. They have no control over their lives and livelihoods, the forest that had once sustained them has all but disappeared; they are exploited by everyone around—the moneylender, the middleman, government functionaries and they are desperately poor.

The small patch of land that they own does not give them food for more than few months a year. They are forced to migrate to far away places in search of work, and when they return, it is only to surrender their earnings to the moneylender with whom they have a life-long covenant of debt.

Jhabua is God's bad country. With the forest cover depleted, the topsoil is lost with each rain. Not that there is much rain either: it plays truant every other year. The groundwater has not been charged for years. How will the land ever yield a crop to feed the Bhil for the entire year.

The Bhil also lack educational and health facilities. They are vulnerable to recurring droughts and dislocation from their homes for long periods of time. The institutions of the state and society treat them shabbily. It has been observed that the Bhil lack power to influence decisions affecting their lives.

These are dimensions of their disempowerment. The deprivations that the Bhil suffer keep them from leading the kind of life that most people take for granted. An escape from this deprivation seems almost impossible and for the Bhil, who have lived a life without them, is a distant, unrealisable dream.

The watershed project in Jhabua seems to have changed all that. Now that the project is almost complete, the results are impressive. A bleak, sandy land has been nursed back to life. Trees are beginning to grow, the soil no longer erodes and the rainwater is conserved.

The Bhil have gained in other ways too. They do not go hungry now like they did in those lean days before the *kharif* crop was harvested. They do not visit the moneylender as often they did earlier; self-help groups give them the credit they need. They do not migrate to distant places in search of work because there is plenty of work in the village itself. There is drinking water, fuel and fodder in the village too; the Bhil women do not have to travel long distances to fetch them.

The Bhil have built and nurtured community organisations that are truly democratic. They have identified the needs of their community and implemented measures to fulfil them. By functioning in user groups, they have ensured that the members of the watershed community have equal access to the benefits that the project offers. The Bhil women have formed self-help groups that ensure thrift and credit to the community. The groups have saved and mobilised enough money to fund the credit needs of their members. Working in these groups, the uneducated Bhil women have managed the complex processes of keeping accounts, charging interest, and networking with financial institutions to increase the capital stock of their groups. They have learnt to recover loans with a tenacity that would be the envy of a well-managed bank.

Have the Bhil found their voice? The answer is yes. The project has put in place an integrated strategy that has given them control over their livelihood. It has mitigated the adverse effects of drought and prevented further ecological degradation. The community organisations created by the project have enabled the Bhil—even the women and poor amongst them—to establish their role in the decision-making process of village development. The Bhil now have a high level of awareness about social issues, a sense of economic self-confidence, and a feeling of self-worth. The project has given them new skills, new knowledge and new efficiencies; they have a voice in the institutions that shape their life.

True, the watershed project in Jhabua is a government project. But it is not the typical top-down, run-of-the-mill government programme administered by crusty bureaucrats with a hand in the till. It is a people's programme and perhaps that is why it has succeeded. The agencies implementing the project have sensitised the Bhil, mobilised them and put them in charge. This is no mean task, given the setting in Jhabua where the Bhil shy away from any contact with the outside world. But the biggest contribution to the project has come from the Bhil themselves. They have contributed substantially to the project: land, traditional knowledge on local conditions, physical labour, savings, and last but not the least, a tremendous degree of commitment.

The story of Jhabua is an amazing one. It is as much a story of ensuring an ecological balance as it is of evolving livelihood strategies for the vulnerable. Above all else, it is a narrative of empowerment—how the women and poor have found a voice in a community that has been marginalised through history.

The book is structured in three parts. The first part deals with the disempowerment of the Bhil from perspectives of history, sociology and environment. The second part looks at the empowerment process of the Bhil as a result of the implementation of the watershed project and the third part contains the conclusions of the book.

The book examines a conceptual framework of empowerment through which the contents of the project and the action plans are analysed. Besides recording interactions with the stakeholders—the civil society intermediaries and the Bhil beneficiaries—the book also analyses and assesses the performance of the project.

Part-I

1

A Historical Profile

It all began with the ancient scriptures[1] wherein they use the word 'Nishada' to describe the Bhil. The Mahabharata traces the origin of the Bhil from the thigh of the sage Vena, the son of Asga and a descendant of Manu Swayambhu, who was issueless. According to mythological sources, the sage rubbed his thigh and produced a short man like a charred log with a flat nose, who was told to sit down—*nishad*—and therefore known as Nishada. From him sprang the Nishada dwelling in the Vindhya mountains, and they were notorious for their wicked deeds.

Another version of the Vena story talks more directly of the descent of the Bhil. In this account, sage Vena was so tainted with sin that the *rishis* decided to pay him a visit asking him to reform, but he was rude to them, i.e., he imperiously waved his hand and asked them to leave. This incensed Angira, one of the *rishis* who then cursed him, and Vena's offending hand was turned into a churning stick, from which sprang the Nishada. When he began to churn with his left hand, three more men emerged: Mushahantara, Kolla and Villa. These were the first ancestors of the Mushaharas, Kols and Bhils (Roy 1970: 34).

There are other legends that speak of the Bhil. One such story revolves around Lord Shiva (Varma 1978: 3). Five men went to see Lord Shiva. Parvati, his consort, told her husband that the men were her brothers, and they had come for the bride price. Lord Shiva gave them a sumptuous feast.

'Tell me,' Lord Shiva asked them after the meal. 'Have you come to collect the bride price?'

'Yes,' the men said.

'Don't you know that I own nothing?' Lord Shiva explained. 'Except Nandi, the bull and this Kamandalu—the pot to carry water, what can I possibly give you?'

The men left but Lord Shiva wanted to give them something. So, he put a silver stool in their path. The men didn't see the gift and went home. Parvati knowing what had happened sent for her brothers and told them,

'You see, my husband was planning to give you a silver stool. He placed it in your path. But you're so stupid that you didn't even notice the stool.'

'Eh! the men said.

'Do you know something, my brothers? There's very little hope of your prospering. But let me do something for you.'

'What'll you do?'

'Well, I'll think of something. By the way, did you see that bull Nandi when you came here?'

'Yes,' the men said.

'Listen! That bull can give you great wealth.'

The men went home and started talking about the bull. The eldest brother suggested that they should kill the bull and take the wealth. The other brothers didn't like the idea, but the eldest brother was so insistent that they finally gave in. They killed the bull, but there was no sign of wealth.

Parvati called them and told them, 'You are all fools, aren't you?'

'Eh?' they said.

'You should have yoked the bull to the plough and reaped wealth from Mother Earth.'

'Oh, we didn't think of that! they wailed.

'You people are very stupid,' Parvati said. 'That was a horrible thing to do—killing a sacred animal out of greed! For doing this, you'll lead a miserable existence and I'll never again look at your faces.'

The characterisation of the Bhil is complete. In Lord Shiva's story, the Bhil is condemned to a miserable existence because he killed a bull, a sacred animal; as Eklavya he is not allowed to excel in archery because he is a Bhil; as a Nishada inhabiting the Vindhya mountains, he is notorious for his wicked deeds.

According to Bhil legend, they were the original owners of the land and were once the ruling race in Rajputana, central India and Gujarat. But they were subjugated by the Rajputs in the 5th century. This could be true because the Rajputs recognise that the Bhil were the original owners of the land. Whenever a new Rajput king is crowned, his brow is marked with blood drawn from a Bhil. At the coronation of the new king, the Bhil place the blood marks of their thumbs on the forehead of the new king, the idea being that the Rajput chief gets admitted by a covenant of blood to the kin of the ancient rulers of the land. This is true, even today, in the old Rajput states of Dungarpur, Kota, Banswara and Pratapgarh (Tod 1829: 209).

The Bhil ranked very high in the affections of the Rajput rulers as they were loyal supporters of the Ranas of Mewar. During the Battle of Haldighati, when Rana Pratap was forced to seek refuge in the forest, it was the Bhil who protected the members of the Rana's family from the Mughals. They carried them in wicker baskets and hid them in the tin mines of Jawaura where they guarded and looked after them. When the family of Rana Pratap stayed in Jawaura, these wicker baskets were hung in the trees and used as cradles for the royal children of Mewar. Colonel Todd writes of how the bolts and rings with which the baskets were hung were still preserved in the trees around Jawaura (Todd 1829: 370). For this act of support, the Ranas of Mewar were ever grateful; they honoured the Bhil by giving them the pride of place in the state emblem of Mewar that has Rana Pratap and a Bhil standing on either side of Eklingji.

At this point in history, the relationship of the Bhil with the Rajputs was very cordial. Marriages between them were fairly common, particularly with the families of the Bhil Chiefs. This created a new caste—the Bhilala—consisting of the descendants of the mixed Rajput and Bhil marriages. There were also cases wherein the children born of a Rajput and Bhil union were recognised as Rajputs. Back then, the norm of eating and drinking with the Bhil was permissible for the Rajputs and all the other castes also took water from the Bhil.

But these mores changed when Hinduism grew more orthodox in Rajputana, and that was when the Bhil were treated as outcastes. So, even though the Rajputs were helped by the Bhil on numerous occasions, they treated them as beasts, particularly in the later period of Rajputana sovereignty. Matters took a turn for the worse when the Rajputs regarded the Bhil as sub-humans and meted out cruel treatment to them. The Bhil were routinely tortured by the Rajputs: they were whipped and beaten with shoes; their eyes gouged out; they were hung from trees and their legs were cut off (Naik 1956: 19).

Perhaps such intense indignity at the hands of the Rajputs drove some Bhil to rebel—particularly dacoity and highway robbery. No doubt, knowing the hilly terrain like the back of his palm aided the Bhil. Catching them in that terrain was not an easy task; therefore, it is no coincidence that, in the 16th century, we find references to the Bhil as freebooters, terrorising the countryside in the Vindhya mountains. Jhabu Naik, who gave Jhabua its name, was one such freebooter operating in the Jhabua region. He was not alone; the south-western districts of Malwa abounded with such freebooters: there was Thana Naik of Thandla and Lakha Naik of Dhulet. Between them, they even killed the son of the Governor of Gujarat.

It was only at the beginning of the 17th century that Keshodas appeared on the scene. A descendant of the Rajput kings of Jodhpur, he served as an apprentice to prince Salim, later Emperor Jahangir. Keshodas distinguished himself in the Mughal campaign of Bengal, and after the accession of Jahangir was given the task of subduing Jhabu Naik. Keshodas managed to tame Jhabu Naik, brought a semblance of order to the Jhabua region; and in recognition of his services, he was granted 10 districts of Malwa. This is how Keshodas came into the possession of Jhabua, Jhabu Naik's territory. The state of Jhabua continued to be ruled by the descendants of Keshodas till the country became independent.

Around 1800, the Scindias and the Holkars started feuding in Khandesh, and there was anarchy in the area. The Bhil organised themselves into groups, and they tried to establish their dominance by taking advantage of the prevailing anarchy. They started highway robbery and lived in bands in the mountains. It appears that the collapse of the system of deploying the soldiers to protect the country aided them indirectly (Varma 1978: 7). The Bhil seized the opportunity: they used to appear in hundreds and attack towns and villages, and carry away hostages for whom they demanded outrageous ransoms.

Then during the reign of Bajee Rao as the Peshwa, Jashwant Rao Holkar instigated the Bhils to destabilise things in the Khandesh region. This is known as the Bhil rebellion of 1809. There was total anarchy and lawlessness. There were at least 50 Bhil bandit leaders with a following of 5,000 Bhils committing acts of plunder and pillage (Sherring 1974: 291–95).

In 1818, Khandesh came under British rule. The Bhil, extremely suspicious of foreign rulers, were not prepared to submit to the regime of order and restraint that the British sought to impose in the region, and rebelled again in their characteristic way. The Satmulla and Ajanta Bhils were in arms this time, terrorising the southern parts of the province, and left behind a trail of blood in the entire range of the Western Ghats. The roads were impassable, neighbouring villages were ruthlessly plundered, murders committed almost on a daily basis, and cattle and hostages were taken away from the very centre of the province (Sherring 1974: 291).

The British adopted two different strategies to deal with the Bhil. The first was to use brutal force. The more refractory of the Bhil leaders were captured, tortured in public and even slaughtered. Others were banished or imprisoned or subjected to public flogging. Coercion was used in full measure while dealing with the leaders, the idea being to set an example.

The other strategy was that of conciliation. The Bhil were offered a lot of sops: pardon was granted to the Bhil that surrendered; land was allotted free of rent; money was doled out for clothes and subsistence, and animals and agricultural implements were provided to the Bhil in order to encourage them to settle down to a sedentary life (Graham 1868: 6).

Initially, the Bhil did not trust the British Government one bit, used as they were to view all outsiders with great suspicion. Eventually the conciliatory measures worked and the British managed to win the confidence of the Bhil. The man responsible for this transformation was Captain Outram. He befriended the Bhil and went about the place unescorted. He indulged the Bhil with feats and entertainment, and they, in turn, were delighted with him.

The turning point came when Captain Outram managed to recruit nine turbulent Bhil. One of them was a notorious plunderer and had, only a short time before, robbed the officer commanding the British detachment which had been sent to capture him. This small group of Bhil became very attached to Captain Outram. For that matter, he was greatly respected by all the Bhil; they admired his acts of courage and valour, particularly the felicity with which he had subdued the turbulent Bhil (Graham 1898: 7).

In 1824, the British established a Bhil police force—the Mewar Bhil Corps—and only the Bhil were recruited. The establishment of the Corps, of course, meant that the Bhil were used to subdue the recalcitrant members of their own tribe and impose order in the Bhil territory. The Bhil Corps fought on the side of the British in the first war of independence of 1857: it also helped the British in checking and supplanting local outbursts in the Khandesh area. A second Bhil battalion was raised, but disbanded three years later. With the Bhil turning over a new leaf, the Corps had very little to do.

Interestingly the Bhil, during this period of British conciliation, gave up highway robbery and other lawless activities. As Captain

Graham says, 'He (the Bhil) feels a relish for that industry which renders subsistence secure, and life peaceful and happy. He unites with the ryots in the cultivation of those fields which he once ravaged and laid waste; and protects the village, traveller, and the property of government which were formerly the objects of his spoliation. The extensive wilds, which heretofore afforded him cover during his bloody expeditions, are now smiling with fruitful crops, and population, industry, and opulence, are progressing throughout the land. Schools have been introduced for the benefit of the rising generations; and the present youth, inured to labour, and sobered by instruction, have lost the recollection of the state of older times, when, from their insular position the tribe reported vengeance and hatred upon their oppressors' (Graham 1898: 7).

The British, however, lost interest in continuing to pacify the Bhil and let the law take over. The result being that the Bhil lapsed to a life of crimes. The period from 1878 to 1889 saw the emergence of Tantia Bhil, a cult hero for the Bhil, who is often compared to Robin Hood.

Tantia was born in 1844. He was imprisoned for a year in Khandwa jail in 1878, but escaped after three days from the jail. Then he began his career as a leader of dacoits and became notorious throughout India. In 1880, 200 followers of Tantia Bhil were captured, but most of them escaped from Jubbelpore jail and were reunited with Tantia Bhil. He now extended his operations to the Indore State, and the districts of Hosangabad and Ellichpur. The British spared no efforts to capture him, but without success; they even put a reward of Rs 5,000 for information on Tantia.

The lores of Tantia's (popularly known as Uncle Tantia) mass appeal and generosity are part of the Bhil heritage. The story is still recounted of how Tantia beat a Brahmin almost to death to extract Rs 100 and then returned a rupee to him for charity. Tantia began to tire of his life as a fugitive. He paid a large sum of money to various government functionaries who promised to get him a pardon. Finally, he was tricked to a meeting with an officer in the Indore Army who promised him pardon and eventually captured him. He was tried and sentenced to death by hanging in December 1889. When he was taken to the gallows, the Bhil flocked in large numbers to catch a final glimpse of Tantia (Russell 1908: 47–48).

With the passage of time, however, the Bhil gradually started giving up highway robbery and other lawless activities. The subsequent history of the Bhil was one of accord marked by isolation and poverty.

After the country became independent, a movement was launched by the Socialist Party, asking the Bhil to convert the forest land to agriculture that would help in strengthening the Bhil's right to the forest land. The gullible Bhil lost no time in cutting down trees and destroying the forest; the wooded landscape was reduced to bald hillocks which were then used for cultivation. The reprisal from the government was swift. The implementation of various forest laws was tightened. As a result of various punitive measures of the government, there was a great deal of harassment and the Bhil were very resentful. A fresh spurt of wrongdoings and highway robbery was the response. The Bhil crimes and rebellion continued well into several decades afterwards, and the crime rate continues to be high even now. Jhabua has the dubious distinction of being the district with the highest crime rate in the state of Madhya Pradesh. Also, it is commonly perceived that most Bhil commit one heinous crime after another out of sheer habit, and that Jhabua is a district that abounds in hardened criminals.

What kind of crimes do the Bhil commit? Topping the chart is homicide. In their area, on an average 273 homicides are committed every year, which means that almost every 72 hours, a human being is killed (Varma 1978: i). Interestingly, the Bhil kill only other Bhil; very few non-Bhil feature in this list. The murders are usually committed within the family and the assailants mostly close relations or family members. What characterises these killings is that they are done without any premeditation. The killer himself is full of remorse after he has dealt the fatal blow with his *phalia* (short, curved knife) or arrow as he/she cannot comprehend why exactly he/she acted in the manner that he/she did and killed someone he/she knew. Moreover, most Bhil who commit murders seem unconcerned about the *prima facie* evidence (in this case, eyewitnesses) that they leave behind on the scene of the crime (Varma 1978: 329, 351).

The behaviour of the Bhil, after he has killed, is unfailingly predictable: he kills and then absconds. But he is often quickly apprehended, as the Bhil are not habitual or professional criminals and cannot cope with the tension that comes from being a fugitive. Moreover, the Bhil community is not one that is inclined to provide shelter or render any help to a law breaker; in fact, his family not only helps in apprehending him but also routinely depose against him in the court of law. The news of homicide spreads like wildfire in the neighbourhood as the Bhil villages are small settlements. Since the general attitude of the community is not to shield an offender, the Bhil are, almost always, keen to apprehend the criminal.

Some of the killings are related to property disputes. Strangely, these killings are committed more out of anger and resentment than a mere desire for possession. This notion can be substantiated by the observation that most Bhil have very few possessions—a small hut, a piece of non-irrigated land, oxen, bow and arrows, a small store of maize and millets, a few metal utensils, and perhaps some coins and currency notes stashed away somewhere.

Other crimes include cattle theft (Varma 1978: 126) and kidnapping; the latter, although severely condemned, often constitutes the necessary step to matrimony in the Bhil discursive practice. This custom will be dealt at length later in this book.

It must be noted that the driving force behind most acts of pillage and roadblocks committed by some Bhil is mere excitement and not material gains. This can be illustrated by two examples. Bhil gangs looted two passenger trains on December 19, 1973 and July 20, 1975 (Varma 1978: 343–44). Armed with their bows and arrows, they stopped the train, looted some passengers, and took away bales of cloth and vegetables from the parcel van. In the first hold-up, the booty was worth Rs 30,000 and in the second, less than Rs 2,500. Out of the booty of the first raid, property worth Rs 14,000 was just discarded soon after the robbery.

For Bhil boys who shoot arrows at public transports or trees, it is just a matter of sport. These boys are natural recruits to the Bhil gangs when they grow up. However, the Bhil, in general, do not see these gang activities as reprehensible.

Almost always, it is observed that most crimes are committed under the influence of liquor. Interestingly, if one were to analyse the defence put up by the Bhil in the courts, although it is common knowledge, not a single crime is attributed to drunkenness.

The Bhil's fondness for liquor is legendary, and it is consumed in prodigious quantities at all special occasions and ceremonies. Liquor is an inseparable part of the Bhil tradition; the gods, the Mother Earth, the ancestors and even evil spirits are propitiated with liquor; of course, it is also the cause of numerous tragedies (Varma 1978: 334).

Although crimes committed by some Bhil are never attributed to drunkenness in courts, the fact remains that toddy juice or Mahua liquor does raise the spirit of a Bhil and makes him bold. Also, it is observed that most misdeeds are committed in the toddy and Mahua months of February, March, April and May: March, April and May are months when the toddy trees start yielding juice and in late April, the Mahua flowers blossom.

Two kinds of explanations (Varma 1978: 77–78) that have been offered for the misdeeds committed by the Bhil include volatile temperament and genes. The former has often been the basis of verdicts by the courts. Although cases of homicides are commonplace amongst the Bhil, not a single culprit in any of the cases has been sentenced to death; even shockingly brutal murders have only been awarded life time imprisonment. More often than not, the courts have justified such leniency on the grounds that the accused, a Bhil, does not fully comprehend the consequences of his action; such justification has also been extended to very few cases of premeditated killings.

The explanation of genetics stems from the popular impression that the Bhil are cruel and barbarous. After all, Jhabu Naik who gave Jhabua its name, Thana Naik of Thandla and Lakha Naik of Dhulet were turbulent freebooters who struck terror in the Mughal hearts with their acts of pillage and plunder. At the time of the Bhil rebellion of 1809, the Bhil gangs had terrorised the entire Khandesh countryside with looting and highway robbery. Wasn't Tantia, the cult hero of the Bhils, a freebooter too?

There have been Bhil rebellions in different historical periods, but they need to be put in perspective. As noted earlier, the crimes precipitated when they were treated like beasts by the Rajputs. Then it seems that Jhabu Naik, Thana Naik and Lakha Naik were merely responding to the indignities heaped upon them by the Mughals. The provocation in most cases was, admittedly, extreme and the Bhil responded in the only way they knew they could. But one should not forget the period when they were treated like humans and they became model citizens. Didn't they lead a crime-free life and display exemplary civility when Captain Outram extended the olive branch to them? Obviously, the explanation for the crimes committed by the Bhil does not lie in their genes, they have to be located elsewhere.

It may make sense to position it in the marginalisation of the Bhil through the centuries. Throughout their troubled history, the Bhil's lack of voice in the processes that have shaped their lives need to be taken into consideration while talking about the Bhil rebellions. We need to note that they have been exploited by everyone around —the Rajputs, the Mughals, the British, the present governments, the moneylender and the government functionaries. The forest that once belonged to them and where they lived without any outside interference was taken away and all kinds of restrictions now bedevil them. The usurpers exploit them in many ways—usurious money-lending, fraudulent dealings and even outright cheating.

Consequently, the Bhil have developed a kind of cynicism that places a very low premium on the value of human life: the Bhil do not mourn the loss of life as seriously as other people. If they had to devote their time and energy to mourn every death that they themselves have caused, there would be no end to their angst and remorse. So, perhaps, the Bhil rebellion needs to be viewed against the background of such cynicism which is but a manifestation of the various dimensions of the Bhil community's disempowerment over time. It is a pity, but the fact remains that the only legacy that the Bhil have inherited in their tumultuous journey through history is the discourse of the marginalised.

Note

1. For more information, see Mahabharata, Visnu Purana and Hari Vamsa.

2

A Sociological Perspective

Customs and rituals

The Bhil are simple people in their dress and in their habits. The men wear a loincloth, a shirt and a turban, which is in fact an inseparable part of the attire. In fact, a Bhil without a turban is the butt of ridicule and reproach. It even plays a role in their customs: when a husband and wife part ways, the man tears off a part of his turban and gives it to her (Varma 1978: 31). The women are equally simply dressed in a skirt, a blouse and *odni*—the unstitched garment that doubles as a waistcloth and headdress.

For the Bhil, clothes are for protection. So poor are they that most Bhil can afford only one set of clothes per year. Both men and women wear earrings and rings made of white metal or brass. Only the wealthy Bhil (of which there are few) wear silver and gold (Lvard 1909: 36).

The Bhil are not too fastidious about doing their hair. The men wear their hair long; it is either partly plaited and fastened with a wooden comb, or allowed to fall in unkempt masses over the shoulders. They generally shave their beard, but keep their moustaches on. The women follow a ritual of applying oil and occasionally washing their hair with curd or milk. They either tie the hair at the back in a bun or wear pigtails.

Tattoos are an integral part of the Bhil ritual. They get it done in the weekly market, and as a special treat at the big fairs. Between the age group of 10 to 12 years old, young girls get tattos on their cheeks, forehead, below the elbows, chin, wrists, and the calf of the leg and feet, while eight or nine-year-old boys got tattoos on their arms, wrists and chests. For the Bhil, tattoo marks bear direct significance to their after-life i.e., it is believed that after death, a Bhil is asked whether he has been pricked by thorns in the forest and is obliged to show the tattoo marks in affirmation. Without the marks, the Bhil would continue to be pricked by thorns in his after-life (Varma 1978: 33). Tattoos of varied designs are done with the help of a needle, and by using the concentrated paste made of bamboo shoot and juice of the Mahua fruit.

The Bhil use earthern vessels for cooking, drinking and eating, these are all bought from the local markets. Aluminium or base-metal vessels or utensils are used only by the affluent Bhils. They eat simple food. Their diet consists of maize, coarse rice, *jowar*, ground-nut and pulses like *urad* and *tur*. Maize, their principal cereal, is often eaten in the form of roasted fresh corn. The women also grind flour from maize every day using the hand-mill. The flour is used to make *roti*, *paniya* or *rabdi*; the latter is a thick gruel consisting of maize flour without milk or sugar, but it is a treat for the Bhil.

Among the pulses, *urad* is more commonly used than *tur*. *Dal*, seasonal fruits like Mahua, mangoes and other jungle fruits are also consumed. Milk and milk products are not very common; their con-sumption depends on the economic status of the family and is mostly restricted to families that rear goats and cows. The Bhil is essentially a non-vegetarian, but he cannot afford to eat meat daily. Often home-grown, meat is mostly eaten during marriages or pro-pitiation rituals.

Drinking liquor is a vital part of the Bhil social tradition; even the women drink toddy or Mahua liquor. It is used for all ceremonies from the birth of a child to the death of an old man. The story goes that if a Bhil is asked what he would ask God to give him, his first choice would be liquor.

Toddy trees are owned by the Bhil. During lean months toddy also doubles up as a food item. Mahua liquor is a local favourite. The entire process of making it, right from collecting the corolla from the Mahua trees in April–May to drying, brewing and distilling the liquor, takes about eight days. The law allows it to be distilled at home for personal consumption; the Bhil are permitted to keep five bottles of Mahua liquor. Liquor is also purchased from the liquor vendors in the area, and shops in the market sell it. A large portion of the household income of the Bhil is spent on liquor (Desai 1996: 1).

The traditional weapon of the Bhil is the bow and arrow. No Bhil house is considered complete without a huge cache of bows and arrows. There are six kinds of arrows to suit the requirement of each occasion: Ghadiyal, Jamni, Bhalka, Tavadia, Kanyali and Bitla.

Ghadiyal is used for shooting at long range: the head of the arrow remains fixed. Jamni has a broad head; it makes a shallow but wider wound, and used to shoot deer. Bhalka, a heavy arrow, is used at short range and even as a spear on tigers. Tavadia is a sharp pointed arrow with four edges. Kanyali is like the Jamni: if it is pulled out, it brings the inner parts with it. Bitla is a blunt button-headed arrow used by beginners to shoot birds (Lvard 1909: 33–34).

The Bhil regard the horse as the sacred animal of the tribal deities. This privilege is not even conferred on the cow. They are also not averse to eating beef. Interestingly, the stone memorials that are erected in honour of the dead in the Bhil community invariably show the dead man riding a horse (Varma 1978: 355). The regard for the horse is so high that even moneylenders, as a form of security from any form of Bhil attack, ride horses in the village. This also explains the success of horse-mounted police in maintaining law and order in the Bhil area.

The Bhil are fond of music and dance. They have a number of musical instruments, the drum being the most popular. The Bhil drum records three main notes: for joy, grief and fear. Other instruments are the tom-tom, cymbals, small and large-sized bamboo flutes, a mouthpiece string which is held tight and vibrated with the nails, a small tambur made from a coconut shell and horse's hair attached to a small stick, and a bronze plate beaten with a stick.

The festivals and fairs are very important for the Bhil (Varma 1978: 36–39). Wavni, that marks the sowing festival, is to propitiate Mother Earth. A *puja* is performed by fixing a plough in the middle of the field, smearing vermilion on the yoke, offering coconuts and pouring liquor on the first furrow. Both men and women participate in this festival and consume liquor. The Divasa festival is celebrated when the maize crop is a foot high. During this festival, the *Pujaro*— the Bhil priest—collects donation from the families: liquor and a goat. He follows the ritual of worshipping the deity, slaughtering the goat and distributing the sacrifice among the participants. Gallons of liquor are consumed during this ritual. In the Nawai festival that takes place when the crop ripens, the new crop is mixed with the old and cooked along with vegetables, which is offered to the ancestors. Being essentially a festival to worship the ancestors, it is taboo to use the new crop before the Nawai festival.

As far as social events go, Bhagoria Hat is the most eagerly awaited event for the youth; it is an occasion for them to choose their spouses. It is held on the market day just before the festival of Holi. On this day, young boys and girls dressed in their finery visit the fair to find a match. According to the custom, the boy pursues the girl he likes and smears her face with red powder and if she reciprocates, then it means that she is willing to be his wife. After a brief period of courtship, the couple returns to the house of the bridegroom and settles down to conjugal life.

Although weddings in the Bhil community are occasions for celebration, the steep bride price (from Rs 10,000 to Rs 35,000) takes a toll on the bride's family (Desai 1996: 11). The secondary status of women, child marriages have taken a toll on the women; the rising mortality rate of teenage mothers in Jhabua district between the age group of 15 to 19 years is evidence of this fact, which will be discussed later in this chapter (*ibid.*: 10).

Structure of the Bhil village

The Bhil community follows the age-old institution of the traditional panchayat. It consists of the *Tadvi*, the *Pujaro* and a few elders of the village. The meeting is usually on a fixed day and time at the house of the *Tadvi*. Liquor is served to the members during the deliberations, and the panchayat remains in session for hours. The traditional panchayat discharges many functions: modification or introduction of customs for ceremonies; settlement of disputes relating to marriages (restoration of brides, adultery) and ancestral property; matters relating to breach of village mortality, and aid being provided to the indigent; and sometimes, cases of homicide are quietly settled by the traditional panchayats; this is so when the intention is not to inform the police (Varma 1978: 42–43).

The traditional panchayat settles a dispute after hearing both the parties and witnesses. The punishment is usually a penalty in cash or a community feast, and in some cases, both. Sometimes, the offender is excommunicated and his re-entry into the village community is possible only after performing the purificatory rites. Strong social ties ensure that the traditional panchayat's decision is all-binding, leaving no room for disobedience. With the advent of the statutory panchayat, the functions of the traditional panchayat have been reduced in scope; they are now limited to social and religious issues, but there is no diminution in its authority. This is because the same coterie of influential people in the Bhil village gets elected to the statutory panchayat; even the women representatives are either from the families of the traditional leaders or affliated to these families.

Tadvi *or village headman*

The most important man in the Bhil village is the Tadvi— the head-man. The Tadvi is a hereditary office and is held by a particular family for generations. When a Tadvi dies, his son becomes the Tadvi even though he may be a minor. Advice of the Tadvi is sought by the Bhil in all matters—settling quarrels, disciplining alcoholics and stubborn husbands. He takes the initiative in propitiating angry gods (in the event of a smallpox epidemic) and he holds an import-ant position in all Bhil festivals. He acts as an intermediary between the government and the Bhil on matters of public concern. Because of his status, the Tadvi wields great influence over the Bhil in his village.

Badwa, *the witch-doctor*

As the Bhil have tremendous belief in witchcraft and magic, the *Badwa*, the witch-doctor, is much in demand in the Bhil community. Should any Bhil fall sick, he is summoned to ward off the evil in-fluences. He is not just a witch-doctor but also the custodian of Bhil theology and mythology.

Pujaro, *the priest*

The *Pujaro*—the priest—is also a practitioner of medicine: he uses roots, leaves or some concoction for curing ailments. He chants *mantra*s to cure the sick. But his main job is to worship the gods on all auspicious occasions. He offers *puja* both on auspicious days and at times of distress. For his services, the Bhil offer some grains to the *Pujaro* at the time of the harvest. He manages the village fairs where he invokes the gods to attend and guard the village. Regarded as an intermediary between the gods and the Bhil, he is revered for this.

Moneylender

Every Bhil village has a moneylender. The moneylenders, originally from the neighbouring states of Rajasthan and Gujarat, have been in the Bhil villages for several generations. As the rate of interest they charge is exorbitant, no Bhil family is ever able to pay back its loans, and the burden of debt continues from father to son in a vicious circle. Much of what the Bhil produces and earns goes to the moneylender. The moneylender is also a trader. He trades in foodgrains, Mahua, forest produce and honey that he buys from the Bhil, and is also a dealer in clothes, ornaments and other items that are sold in the Bhil area. Although they are aware of the fact that they would get less money from transactions with the moneylender, the Bhil continue to take their produce to him.

Ironically, the Bhil regards the moneylender as an ally even though he is almost always in debt. Whenever the Bhil is in need of money or foodgrains or clothes or ornaments, he approaches the moneylender who lends him the items or cash for a thumb impression in the register of the moneylender.

According to a survey, 88 per cent of the Bhil families in Jhabua are indebted to the moneylenders (NIRD 1987). Perhaps no other community in India is more in debt than the Bhil.

Why are the Bhil so indebted? Here are some reasons: For a regular Bhil, liquor eats into the bulk of his earnings. And they also have social obligations: the whole panoply of ceremonies and functions that make huge demands on their meagre income. Marriage and the after-death ritual, the *nukta*, also involve considerable expenses.

The Bhil's amicable relationship with the moneylender comes from the fact that there is no paperwork involved in the dealings as they have a pathological fear of the written word. Most of them are illiterate, and the comfort factor in their dealings with the moneylender is that all they have to do is to put their thumb impression in a register and get what they want.

The moneylender, on the other hand, is aware of his hold on the Bhil, and the way he conducts himself in the Bhil village is ample testimony to this. He can visit even the remotest part of the Bhil territory without the slightest apprehension of any Bhil ever being disrespectful to him.

Government functionaries

The Bhil also shy away from any contact with the outside world. It does not help that they are treated shabbily in whatever little contact they have with the outside world; whether it is the Block Development Officer or the Anganwadi worker or the conductor in the public transport system, they are at the receiving end. This makes them fearful of the government. As the Bhil point out, 'government people always come to take something from us and if we do not give, we will be beaten up or sent to jail or have to pay money or face difficulties' (Desai 1996: 3).

As most Bhil are illiterate, they are ignorant of their rights and entitlements as citizens. Only a handful of people control information in the Bhil community (Desai 1996: 4). And it so happens that they are not eager to part with information, because information in the Bhil setting is synonymous with power. Therefore, it is primarily the lack of information that has led to the exploitation of the Bhil. For instance, they are obliged to pay commission to the middleman to avail of the benefits given by the government. Although the government spends a large amount of money in Jhabua district by way of development schemes, very little reaches the poor Bhil (Sisodia 1996: 31; also see Desai 1996: 15). It is quite common to see the Bhil getting a paltry sum of Rs 800 after putting his thumb impression on loan papers for Rs 2,000. The rest goes to the middleman. When it is time for recovery, the Bhil approaches the moneylender for a loan to pay off the government dues.

Exploitation and Bhil

The exploitation of the Bhil takes place in other ways too. The moneylender and the *Palliwala*—the middleman in foodgrains— buy grains from the Bhil at a very low price when the crop is harvested and then sell them at higher prices in the big towns. The Bhil has implicit faith in the moneylender or the *Palliwala* due to which they sell their grains to them, even without finding out what the market price is or how much they owe to the moneylender. On the whole, the Bhil prefer the moneylender to banks, *badwas* to doctors, and middleman to government functionaries. The patch of land they own does not yield enough to last them for the entire year. They have to migrate in search of work, and this money ultimately goes to the moneylender.

Women

If life is hard for the men, it is harder for the women. Apart from doing the household chores, they share the work-load with the men in the fields; in fact, tasks requiring harder physical labour are performed by women. Malnourished from an early age, heavy labour at their houses and the fields leave them with very little time to take care of their own health. This is compounded by the fact that girls become mothers at the age of 13 or 14. The percentage of girls marrying below 18 years is 58.4 in Jhabua district (Government of Madhya Pradesh 2002: 472). In fact, in the rural areas, the position is much worse: 86 per cent of Bhil women start their reproductive life before they are 18 (Rapid Household Survey 1999: 15) resulting in serious implications on the health of the Bhil women and the children they bear. The upswing of female deaths in the age group of 15–19 years indicates a high rate of mortality among teenage mothers.

It does not help that the Bhil women are uninformed about contraception, child bearing and child rearing. Almost 82 per cent of the deliveries in Jhabua are conducted at home (Rapid Household Survey 1999: 25). When the women have deliveries at home they are attended to by untrained persons (93 per cent). 76 per cent of Bhil women report one or more complications during pregnancy. Of these, only 32 per cent seek treatment (*ibid.*: 25). There is very low awareness of post-partum care. Only one-fifth of the Bhil women avail of antenatal care (ANC) services and of these, only 8 per cent avail of full ANC services consisting of provisions of antenatal care, including at least three antenatal care visits, iron prophylaxis for pregnant and lactating mothers, two doses of tetanus toxoid vaccine, detection and treatment of anaemia in mothers, and management and referral of high-risk pregnancies (*ibid.*: 21). This implies that the general awareness of ANC is very low amongst the Bhil women and compliance, much lower. Although there is no reliable data on malnutrition and anaemia, most Bhil women appear to be anaemic.

In terms of working hours, the women work, on an average, 15 to 20 per cent more than the men. They as a group consume far less than the men but lead a more difficult life. However, controlled cases of domestic violence are not unusual. They are not only marginalised within the household, but are also at the periphery in the social setting, leaving them with no role in the decision-making process or in the management of community institutions. Although the men recognise the worth of the work that the women do, the latter are still relegated to a subordinate position.

Children

Managing a hectic schedule both at home and outside leave the women with very little quality time for child rearing. Very little attention is paid to the health care of the children. The proportion of children in Jhabua who are fully immunised against all the six killer diseases is only 17 per cent (Rapid Household Survey 1999: 29). A number of deaths occur among children under five due to diarrhoea and pneumonia. Oral Rehydration Salt (ORS) therapy is hardly used as a remedial measure against diarrhoea because most mothers are unaware of it.

The process of education for the children starts around the age of five. Boys are taught to make ropes, give fodder and water to the cattle, and tie or untie the bullocks from their posts. The girls stay at home with their mothers and learn cooking and other household chores.

From the age of six, they start assisting and accompanying their elders in all kinds of chores. Children go to the forest to graze cattle. They do household work such as preparing cattle fodder, removing dung and cleaning cattle sheds, and carrying firewood from the forest. Young girls fetch water from the streams and help in the husking and pounding of grains. The report brought out by the National Institute of Rural Development concludes that Jhabua district has the highest percentage of child labour in the country (NIRD 1987). Most children do not go to school, and the ones who do, drop out at the earliest opportunity. The Bhil girls do not go to school because of the long distances they have to negotiate and the insecurity while commuting, and also because of the migration of parents outside Jhabua district in search of work (Singh *et al.* 1996: 3).

Migration

Moving out of the district for a few months every year in search of work is a regular feature of Bhil life. 'My forefathers have done it. I did it and my children have to do it because there is no work in the village'(Desai 1996: 7). The Bhil, once the ruling race in Rajputana, Central India and Gujarat, are nomads now; they are the 'Mama', the contemptuous name given to the migrant labour in cities to which they return time and again in search of livelihood.

The duration of their stay at the place of migration depends on the opportunities in their own village in Jhabua district; i.e., more opportunities in the village leads to less migration. The extent of migration varies, as a recent study shows (DECU 95: 55) on an average, as many as 57 per cent of the Bhil households in Jhabua district migrate in search of work (See Fig. 2.1).

Among the migrating households, 14 per cent migrate *in toto*: the entire family migrates. They distribute their household goods among their neighbours and lock the house. From the other 86 per cent, one or more persons migrate, leaving the house in the care of their elderly people and children. The women migrate with their men. According to the survey, out of the migrating population, 58 per cent are men and 42 per cent women; that means that the proportion of men and women is about 3:2 (ISRO 1995: 56). This also means that the women migrate in sizable numbers; so, even in migration, their contribution to the family income remains substantial.

Fig. 2.1 Percentage of Migrating Households

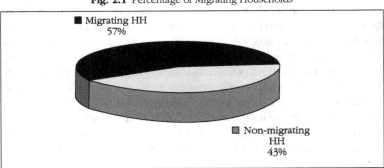

■ Migrating HH
57%

□ Non-migrating
HH
43%

Source: ISRO (1995)

The decision on the place of migration is based on its employment opportunities. A potential migrant is usually in the age bracket of 15 to 50 years with the capacity to work for eight hours a day and six days a week. More than 70 per cent of the migrants are between 21 to 50 years, with about one-fifth of the migrants between the age group of 15 to 20.

Marginal and small farmers account for more than three-fourths of the total migrant population, and as the size of the landholding increases, their representation in the migrant population diminishes. Migration also depends on whether the Bhil household has irrigated land and can raise *rabi* crops in addition to *kharif* crops. If the household can raise *rabi* crop, then the motivation to migrate is decidedly less—it means that the number of days for which work is available in the village is more.

For the migrating population taken as a whole, the average period of migration is 96 days (ISRO 1995: 57). The calendar of migration (See Fig. 2.2) is linked to the agricultural operations. The bulk of it takes place during the months of November–December and during March and April (mostly post-harvesting months after the *kharif* and *rabi* crops). As the chart shows, migration takes place between the months of November and April, and rest of the year they stay at home.

Four-fifths of the migrant households migrate to familiar places (See Fig. 2.3). Kota district in Rajasthan is sought after by migrants, particularly for those from Jhabua and Thandla tehsils. Other popular cities include Surat, Baroda, Dahod and Ahmedabad in Gujarat, and Pitampura, Indore, Bhopal, Mandsaur and Khandwa in Madhya Pradesh. They migrate in all four directions, but as the chart shows, the majority of the Bhil migrate to cities in western India. In fact, 37 per cent go in the western direction while 32 per cent go to the eastern parts of Madhya Pradesh, and 29 per cent to the northern parts of Madhya Pradesh and Rajasthan. Only 2 per cent prefer to migrate to the south; the reason, perhaps, is the natural barrier of the river Narmada.

A common complaint of the migrating families is the harassment by the policemen and the railway staff: They not only insist on checking every item that the Bhil carry but also routinely take away food items by way of bribes. Life is not hassle free for the Bhil at the place of migration. According to the survey, 73 per cent of the migrant families stay on open ground or below trees. Only a few

Fig. 2.2 Calendar of Migration

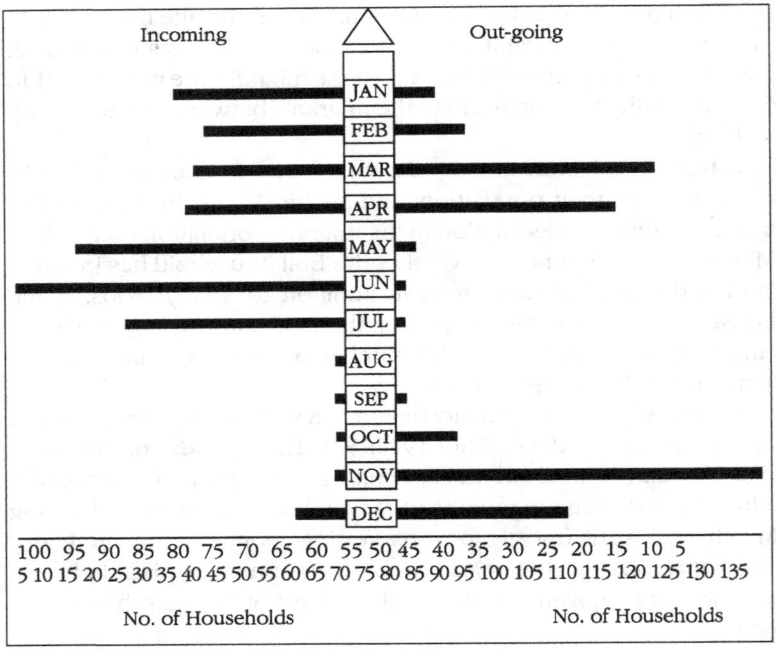

Incoming		Out-going
	JAN	
	FEB	
	MAR	
	APR	
	MAY	
	JUN	
	JUL	
	AUG	
	SEP	
	OCT	
	NOV	
	DEC	

```
100 95 90 85 80 75 70 65 60 55 50 45 40 35 30 25 20 15 10 5
5 10 15 20 25 30 35 40 45 50 55 60 65 70 75 80 85 90 95 100 105 110 115 120 125 130 135
```

No. of Households   No. of Households

Source: ISRO (1995)

of the families (9 per cent) stay in some permanent shelter like hutments, others stay in temples or open barracks or even on footpaths (ISRO 1995: 57). Illnesses are dreaded by the migrant Bhil as medical facilities are not only expensive but also quite inaccessible. According to the survey, 8 per cent of the migrants had to undergo medical treatment.

On an average, the net savings of a migrant family are about Rs 1,227, while for all the households it ranges between Rs 100 to Rs 5,500. It is sad but true that the bulk of this money goes to the money-lender as repayment of old loans.

Migration disrupts the normal life of the Bhil. For the young it is traumatic: they grow up in an unhygienic and relatively hostile environment (Desai 96: 7). The mother works, and has no time to take care of the children. The children, in fact, start working from a very young age at the place of migration.

Fig. 2.3 Migratory Destinations of the Bhil from Jhabua District

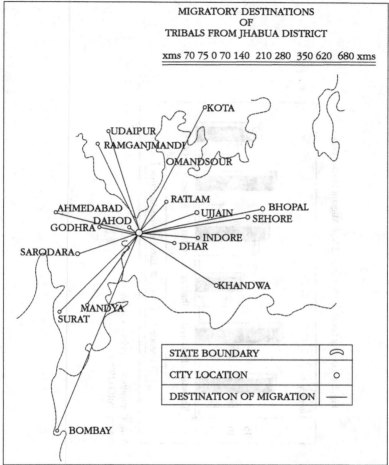

Source: ISRO (1995)

The education of the children is also affected. Those left behind in the villages have to look after the house and animals and they often miss school, while migrant children are not enrolled in a school at the place of migration. These factors have contributed to relatively lower literacy level in the migrant households than in others (See Fig. 2.4).

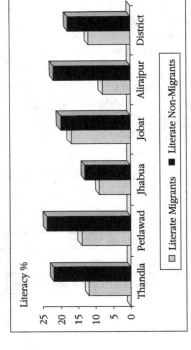

Fig. 2.4 Literacy Level of Migrant and Non-Migrant Households

Note: Computed for persons above the age of 6.
Source: ISRO (1995)

3

Physical Dimensions

In the days when the forest cover was plentiful, the essential needs of the Bhil household—fuel, fodder, shelter and Non-Forest Timber Produce (NFTPs)—were met by it (Krishna Kumar 1997: 6). Much of the forest is gone now. About 76 per cent of the degradation of the forest area in Jhabua district over the last nine decades has been caused by the expansion in agriculture, encroachment for human settlement, commercial felling of trees, cutting of trees for fuel and fodder and unrestricted grazing. A big chunk of the forest area—more than 31 per cent—has now been converted to barren land (Uppal 1996: 19). Loss of forest cover is the most serious environmental threat that Jhabua faces; the result being deterioration of land by erosion, desertification and loss of biodiversity.

The crop area of the district is a little more than half its geographical area. Since much of Jhabua's crop area is converted from forests, the land is infested with roots and shoots of various species and the gradient is high (Fig. 3.1).

Fig. 3.1 Land Use Classification

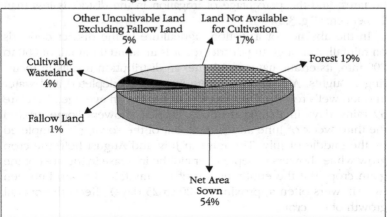

Other Uncultivable Land Excluding Fallow Land 5%

Land Not Available for Cultivation 17%

Forest 19%

Cultivable Wasteland 4%

Fallow Land 1%

Net Area Sown 54%

Source: Government of Madhya Pradesh (2002)

Land here is shallow, sandy, stony and infertile with a soil depth of only seven to 15 centimetres. The soil is poor in its nitrogen content, with only 8 per cent of clay (Uppal 1996: 21).

Soil erosion is a serious problem. Degradation of forest has left behind 36 per cent of the arable land with practically no soil cover. The two types of erosion that plague Jhabua are: sheet erosion (the surface layer of the soil is washed away) and rill or shoestring erosion, which takes the form of a series of more or less parallel small gullies and when rill erosion continues over a period of time, it becomes gullies erosion. Land in the northern parts of the district is affected by sheet erosion while the entire southern part of Jhabua district is affected by rill erosion (Uppal 1996: 22).

All one sees in Jhabua is sandy, steep hill-slopes, albeit barren with no topsoil and therefore no water-holding capacity. Land in most villages is covered in coarse, shallow soil with patches of black cotton soil. There are four types of soil in Jhabua district—the medium black, red, mixed, and alluvial.

Three rivers—Mahi, Anas and Narmada—form the water catchments in the district. Mahi has the highest catchment area with about 20,000 hectares; its basin covers the northern part of the district and almost 54 per cent of the geographical area. Anas covers the central part of the district before it turns away to Gujarat with a catchment area of about 7,800 hectares. Narmada, which delineates the southern boundary of the district, has a catchment area of about 500 hectares (Uppal 1996: 96). The total irrigated area is a mere 80,000 hectares, and the percentage of irrigation in the district is less than 20 per cent (Fig. 3.2).

In the absence of irrigation, agriculture in the district depends on rainfall. Although the district records an annual rainfall of 600 to 800 mm, its erratic nature and uneven distribution results in recurring droughts. At this time, groundwater gets depleted and water in open wells recedes by 30 to 40 metres. On an average, there are 59 rainy days in Jhabua district. The first showers come around the third week of June and 50 per cent of the sowing is completed by the middle of July. The rains in July and August help the crop grow while showers in September aid the increase in the area of the gram crop. But the erratic nature of the rainfall (the gap between two showers often approximates 20 to 25 days) affects the overall growth of the crops.

Fig. 3.2 Percentage of Irrigated Area by Source (Madhya Pradesh, India and Jhabua)

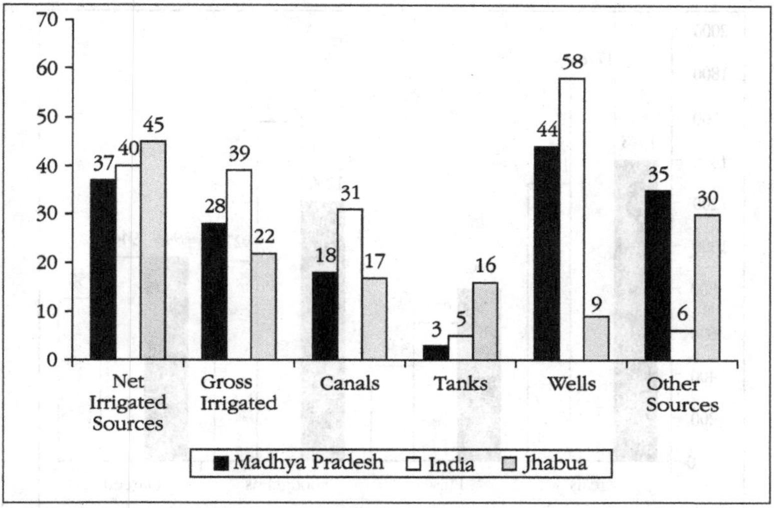

Source: Government of Madhya Pradesh (2002)

Cropping

Crops grown in the district are *jowar*, wheat, paddy, gram and groundnut. Among cereals, almost 50 per cent of the gross sown area in the district is covered under maize, *jowar* and other minor millet. Pulses comprise 30 per cent of the gross cropped area; *urad* is grown in the largest area followed by *tur*. Most Bhil adopt the mixed cropping pattern and cultivate maize with *tur* and *urad*. In the *rabi* season, gram is the favoured crop; it is cultivated in all the blocks of the district. Groundnut is the most preferred crop in oilseeds (Uppal 1996: 96).

The area under the *kharif* crop is about 3,50,600 hectares while the area under the *rabi* crop is about 1,03,600 hectares (Government of Madhya Pradesh 2002). On the whole, the rainfall and moisture retention nature of the soil in Jhabua district is not suited for raising two sequential crops in a year. The average yield of crops in Jhabua compares rather poorly with the national average and the average for the state of Madhya Pradesh (see Fig. 3.3).

Fig. 3.3 Yield of Crops (Madhya Pradesh, India and Jhabua)

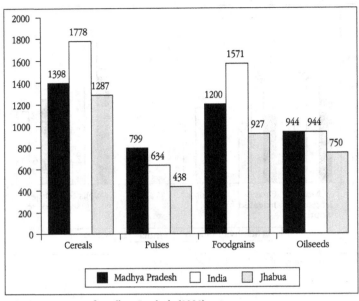

Source: Government of Madhya Pradesh (2002)

The low yield is due to low fertility of the land, low investment, and non-availability of early maturing varieties. Due to recurring droughts, the Bhil choose drought-resistant, local crop varieties that are low-yielding. The farming practices of the Bhil are traditional. These include using the wooden plough; sowing by the old-fashioned drill connected to a plough; bullock-drawn hoes for soil mulching; and manually weeding the soil.

Although the government has tried to provide the Bhil with better implements and more varieties in crops, they have spurned these for various reasons. For instance, the Bhil do not favour the moulded iron plough at a subsidy of 50 per cent as they do not require a plough with a long blade, which is effectively used for deep ploughing. Another reason why they do not favour it, even with the subsidy thrown in, is because it is too expensive. As for the variety of crops, they prefer the *deshi* variety of maize to the *safeda* variety (with a high yield) introduced by the Agriculture Department. Despite looking *deshi* in colour, the *safeda* variety is yet to find acceptance with the Bhil (Uppal 1996: 96).

Landholding

The total number of landholders in Jhabua district is 1.44 lakh and the average size of landholding is 2.23 hectares. But there has been a decrease in the average size of landholdings. While in 1980–81 it was 3.8 hectares, it came down to 3.1 hectares in 1985–86. During 1990–91 it came down further to 2.6 hectares, while at present, it is 2.23 hectares. Marginal farmers in the district who hold less than one hectare of land form about 27 per cent of total landholding, and small cultivators who have land between one to two hectares, 28 per cent. The number of semi-medium farmers, who hold between two to four hectares, is significant at 38,521.

As the size of the holding increases, the number of cultivators decreases. The number of medium farmers who hold more than four hectares but less than 10 hectares is 24,497, and the number of farmers who have more than 10 hectares is only 3,064. The big farmers are found only in two blocks—Sondhwa and Alirajpur—in the southern-most part of the district bordering the river Narmada. The landholding pattern shows a concentration of farming households among the small and marginal farmers. There is an explanation for this. As soon as the son gets married, the family's land is divided and the son's share in the land is handed over to him. Thus, breaking down the land into smaller sizes explains the increasing number of small and marginal farmers in Jhabua district.

Population

The population of Jhabua district has risen sharply over the last several decades (see Fig. 3.4).

Interestingly, while it took 50 years (1901 to 1951) for the population of the district to double itself in the years before independence, it took only 30 years (1961 to 1991) for the population to double itself after independence (ISRO 1995: 33–34). The growth in population, however, has not perturbed the predominantly tribal character of the district (see Fig. 3.5).

There has been an increase of more than 122 per cent in the Bhil population of Jhabua district during the last three decades. The increase in population has created enormous pressure on land, and with the population of doubling itself in 30 years, the density has

Fig 3.4 Population of Jhabua District

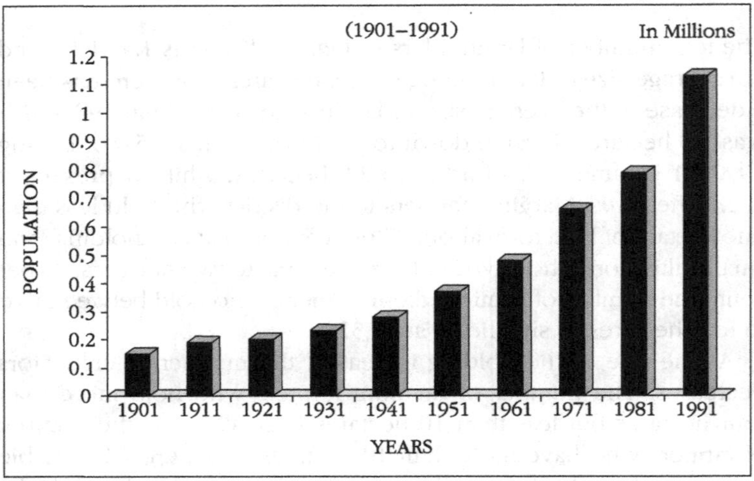

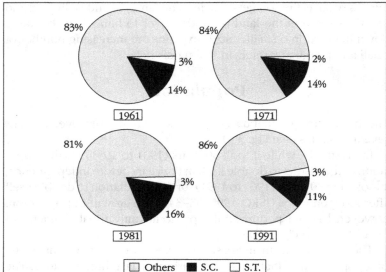

Source: ISRO (1995)

Fig 3.5 Proportion of S.C and S.T. in Total Population of Jhabua District (1961–91)

Source: ISRO (1996)

increased from 76 to 166 per sq. km. Since the rate of mutation in land-holding is linked to the population growth, the implications of the present trend are ominous. In Jhabua district, the number of small and marginal holders has increased while the number of big farmers has decreased. This means that fragmentation is already taking place at a high rate, and since the Bhil population is increasing, it would mean further fragmentation in the landholding of the Bhil.

The increasing mutation of landholding combined with low fertility of the soil affects the agricultural production adversely. There is little possibility of evolving an alternative land use pattern because of the small and marginal size of the holdings. Only the medium and big farmers can afford to take the initiative, but their number is diminishing. This has further implications on the settlement pattern as well. In Jhabua district, a village is spread over a large area and the houses of the Bhil are set apart. They build their houses close to their fields. As the number of Bhil houses grows with the number of family members, the area develops into a *falia* (hamlet). The number of *falia*s varies from village to village; on an average, there are five to eight *falia*s in a Bhil village. Over time, these *falia*s take the form of a village with 25 to 30 houses. The average population in a village in Jhabua district has increased from 360 in 1961 to 551 in 1991 (ISRO 1995: 95).

Employment

While the Bhil population and the number of *falia*s are on the increase, the occupational pattern of the Bhil remains the same. About 73.3 per cent of the households in Jhabua district are cultivators, and 13.9 work as agricultural labourers. Agriculture is their sole livelihood, followed by livestock rearing; the latter supplements food, fuel, fertiliser and finances of the Bhil. Milching cows, buffaloes and ploughing bulls constitute the major proportion of livestock. There are 2,13,813 milch animals (0.16 per capita) and 3,26,466 draught animals (1.93 draught animals per operational holding) in Jhabua district (Government of Madhya Pradesh 2002: 405). The indigenous breed of cattle favoured by the Bhil is undernourished and poor in terms of milk produce and draught quality. The health and general living conditions of the livestock are abysmal. Minor injuries, infertility, and diseases like Foot and Mouth, Haemorrhagic Septicaemia and Black Quarter are common problems.

The quality of the fodder collected in the form of maize, *jowar, urad* and wheat straw along with dry grass is low-grade. The bulk of the traditional grassland in the district is under unauthorised cultivation for agriculture, and green fodder like Barseem is cultivated in a very small area. The fodder deficit is a serious problem as there are as many domestic animals in the district as human beings. It is estimated at 1.27 million tonnes, which would call for a reduction in the livestock population by 61.5 per cent (NCHSE 1993: 93). Dairy-ing is pursued more as a source of household consumption than as a commercial enterprise. Goats are reared in large numbers. Poultry is also a major occupation for the Bhil. The local breed, *kadaknath*, incidentally a great favourite of the Bhil, is reared only in Jhabua district.

Due to the scattered nature of the settlement pattern, there are no artisans or household industry among the Bhil. Their houses are so spread out that the members of a household are skilled in a number of jobs including carpentry, masonry, pottery and tile-making. In fact, if one walks through the *falias*, one notices the absence of small shops and workplace generally associated with artisans (see Fig. 3.6).

Fig. 3.6 Share of Workers in Different Categories (Madhya Pradesh, India, Jhabua)

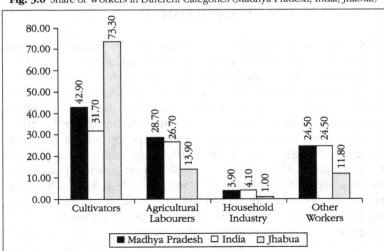

Source: Government of Madhya Pradesh (2002)

There is hardly any organised employment in the district: the total employment in registered industries is negligible at 4,000.

Work participation rate

The Work Participation Rate (WPRs), which shows the total number of main and marginal workers to the total population, gives us important insights into the employment in the district. The census records a person as a main worker if he/she has worked for a major part of the year preceding the enumeration. A person is recorded as a marginal worker if he/she has worked any time at all in the year preceding the enumeration, but has not worked for a major part of the year. On the whole, the Work Participation Rate provides an idea about the participation of the population in economic activity (see Fig. 3.7).

What is striking about the relative position of the Work Participation Rate for India, Madhya Pradesh and Jhabua is that there is a great

Fig. 3.7 Work Participation Rate (Madhya Pradesh, India and Jhabua)

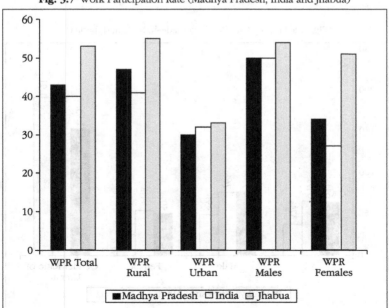

Source: Government of Madhya Pradesh (2002)

deal of difference between the Work Participation Rate of India and Madhya Pradesh on the one hand, and Jhabua district on the other. The over 12 per cent difference between the rural WPRs shows the need for large number of people to be involved in the workforce in Jhabua. This quite clearly shows that the remuneration from most livelihoods in Jhabua is low, and more people in the households are required to work to maintain themselves.

The number of children working in Jhabua district is exceptionally high. Children, both as main and marginal workers, account for 25.5 per cent. This is in comparison to 17 per cent of children for the state of Madhya Pradesh, and that too, the rate of child workers in Madhya Pradesh is the fifth highest among the major states in India (Government of Madhya Pradesh 2002: 211).

Education

Most Bhil are illiterate. The literacy rate in Jhabua district is low compared to the rate for the country and the state of Madhya Pradesh (see Fig. 3.8).

Fig. 3.8 Literacy Rate (Madhya Pradesh, India and Jhabua)

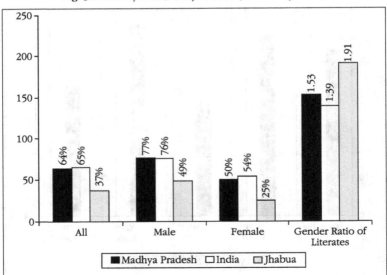

Source: Government of Madhya Pradesh (2002)

The low level of female literacy is of serious concern. To the Bhil, education of girls is relatively unimportant. Girls go to school only until they reach puberty. This pattern is a result of the prevalent social norm that it is useless to educate a girl as she would eventually get married and be a part of another household. This has been compounded by cases of educated Bhil girls refusing to marry uneducated Bhil boys. In such cases, the educated Bhil girls have to be forcibly married off at a higher bride price (educated girls command a higher bride price).

The district has 1,313 villages and there are 2,016 primary schools. Every village has a school; this, ostensibly, satisfies the government norms for opening a school in a village with a population between 250 and 300. But such norms ignore the scattered and inaccessible habitation pattern of Jhabua district where the Bhil live dispersed in small household communities, one to six kilometres away from the main village. This is one instance where the peculiar needs of the Bhil community are concealed within the standardised norms and strategies of the government. The total number of *falias* in Jhabua district is about 8,000. Each *falia* does not have a school and the children have to walk long distances to go to school. The numbers speak for themselves: only 36 per cent of the children in Jhabua district have access to primary school education within a radius of one kilometre.

In most schools, the attendance rate is low. This is particularly so for the girl students. It does not help that a large number of schools are single-teacher schools, and for a variety of reasons, the schools do not function on all the days of the academic session. There are substantial chinks in the educational infrastructure: 416 primary schools and 318 middle schools without any building; 881 primary schools and 267 middle schools in want of additional rooms. 916 primary schools and 60 middle schools without toilet facilities; 600 primary schools and 201 middle schools without provisions for drinking water (Government of Madhya Pradesh 2002: 39–49).

The enrolment data of the district do not present a happy picture. In the age group of six to 14, the enrolment was 2.61 lakh out of which 1.55 lakh were boys and 1.06 lakh were girls, constituting an overall enrolment rate of 69.7 per cent. However, there was a sizable population of Bhil children who were out of school: 18,068 dropped out of school (10,098 boys and 7,970 girls) and 83,478 (37,771 boys and 45,707 girls) never got enrolled (Government of Madhya Pradesh 2002: 46).

Non-formal educational institutions at the pre-primary level—Anganwadis and Poriyawadis—exist in almost all the villages. There are 1,959 Anganwadi centres and 1,508 Poriyawadi centres in Jhabua district but the management of these centres is with the village Sarpanch, and there is hardly any public involvement. The functioning of these centres is dismal.

On the whole, the educational development of the Bhil is dismal. The reasons are: parents' lack of awareness; history of alcoholism in the family; distance between residence and school; absence of community participation in the school management; non-local teachers; schoolteachers unfamiliar with the local dialect; migration; employment of children in agricultural and household work; lack of enthusiasm for girls' education for reasons of insecurity and social norms.

Health

Life expectancy at birth is the most comprehensive indicator of health; a better health status can be safely assumed to give a better life expectancy. The following is a comparative picture of life expectancy in Jhabua district (see Fig. 3.9).

Fig. 3.9 Life Expectancy Birth (India, Madhya Pradesh and Jhabua)

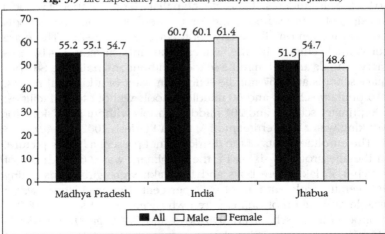

Source: Government of Madhya Pradesh (2002)

The state of infant and child mortality rate is perhaps the best assessment of the basic healthcare, quality and reach of health delivery, general environment for health, crucial health determinants including nutrition, sanitation, safe drinking water, etc. Infant Mortality Rate (IMR) is the number of deaths of infants less than one year of age per 1,000 live births, while Child Mortality Rate (CMR) is the number of deaths among children aged one to five years per 1,000 children in the same age group.

Jhabua's Infant Mortality Rate at 130 per 1,000 is high, compared to 89.5 for the state of Madhya Pradesh and 70 for India (see Fig. 3.10), although it seems slightly better off when it comes to Child Mortality Rate (see Fig. 3.11). But on the whole, child care is dismal in Jhabua district. Only 17 per cent of the Bhil children are fully immunised. 38 per cent of the children receive no vaccination at all (Rapid Household Survey 1999: 29). Bhil mothers are unaware of the remedial measures in case of killer diseases like diarrhoea and pneumonia.

Healthcare amongst the Bhil is deplorable. Ironically, 90 per cent of the deaths in Jhabua district are caused by easily preventable or otherwise curable diseases. Negligence, ignorance and superstition are the culprits. Diarrhoea deaths account for 30 to 40 per cent

Fig. 3.10 Infant Mortality Rate (India, Madhya Pradesh and Jhabua)

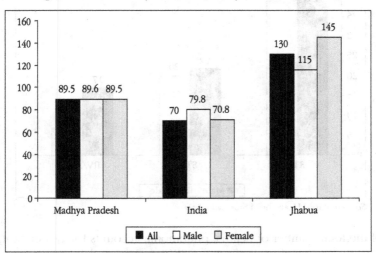

Source: Government of Madhya Pradesh (2002)

Fig. 3.11 Child Mortality Rate (India, Madhya Pradesh and Jhabua)

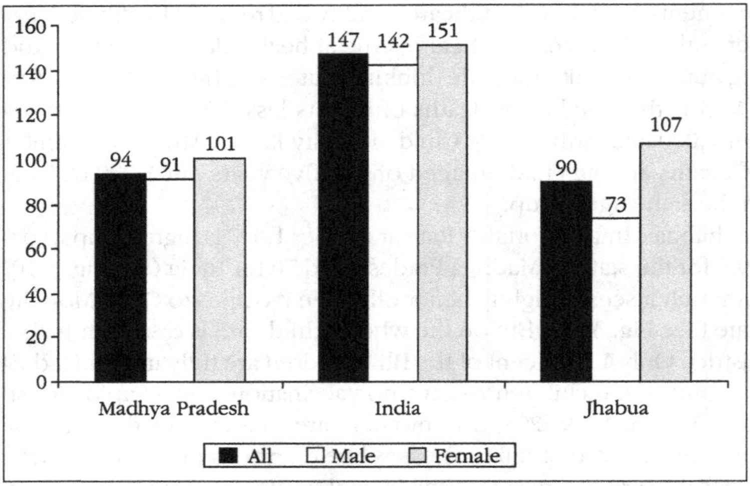

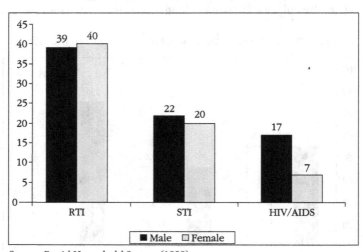

Source: Third Human Development Report, Madhya Pradesh (2002)

Fig. 3.12 Awareness of RTI, STI and HIV/AIDS

Source: Rapid Household Survey (1999)

of the total number of deaths, pneumonia accounts for 28 per cent, vaccine preventable diseases for 28 per cent, and other diseases, for 4 per cent.

The majority of the Bhil still patronise the *Badwa*. The health status of the Bhil has suffered because of the inability of the traditional medical system to cope with the advent of modern diseases and the lack of access to advanced medical care. Also, most Bhil are unaware of killer diseases. For example, the overall awareness of three diseases—Reproductive Tract Infection (RTI), Sexually Transmitted Infection (STI), and HIV/AIDS—is very poor (see Fig. 3.12).

It does not help that unhealthy conditions continue to persist. The number of habitations without potable water is as high as 150. The percentage of population with proper sanitation facilities is a mere 7 per cent in the rural areas: it means that almost all the rural households (93 per cent) use the fields for defecation (Rapid Household Survey 1999: 11). Only 31 per cent of the population have access to Integrated Child Development Scheme (ICDS) services within a radius of one kilometre.

Panchayat institutions

In Madhya Pradesh, a Gram Panchayat is the elected body for a population of 1,000. The main responsibility of the panchayat is to plan the village's economic development and ensure social justice. For its effective functioning it has been given functional control of those departments of the government that need to be locally managed, and the staff and budget of those departments have been transferred to the panchayats.

In Madhya Pradesh, the Gram Sabha—the village community that elects the office-bearers of the Panchayat—has been given several powers and functions. They are:

- Exercise control over government functionaries;
- Undertake planning;
- Identify/approve beneficiaries under different government schemes;
- Provide approval to proposals before they can be implemented in the villages;
- Control and manage various programmes of the government;
- Cooperate with other agencies for the implementation of their activities;
- Undertake the review, monitoring, inspection and supervision of government programmes in the village;

- Maintain records and accounts at the village level;
- Maintain structures created under various schemes;
- Perform a facilitative and promotional role towards the development of the village.

The panchayat system institutionalises an interactive mode wherein people make plans for village development and execute them. As an institution, it is designed to be both representative and enabling: representative in that it expresses local concerns for human development and undertakes consequent action, and enabling in that it creates opportunities for self-governance.

In the Bhil area, the panchayat system has failed in both these roles for two reasons. One is the almost overwhelming influence of the Sarpanch over the panchayat, and the other being that the leadership of the panchayat institutions in Jhabua district has been captured by the traditional leaders of the Bhil community. Even the women representatives, who are elected to the panchayats by virtue of the statutory reservation for women, are either from the families of the traditional leaders or related to these families. The result is that the panchayat institutions in Jhabua district, though technically representative in character, have degenerated into non-participative, authoritarian structures—much like the traditional Bhil panchayats—which are manipulated by the Sarpanch and his cronies.

The Gram Sabha has not been in a position to establish its role, character and importance in the Bhil area. This is for several reasons.

- While Gram Sabha meetings are held on the dates specified by the government—26 January, 14 April, 15 August and 2 October—the attendance is poor. Typically, the attendance ranges from a dozen to two dozen Bhil.
- Gram Sabha members close to or supporters of the Sarpanch dominate the proceedings, thereby denying the Gram Sabha the characteristics of a democratic body.
- The issues discussed in Gram Sabha meetings are matters such as the budget, which are either incomprehensible or of little utility to the ordinary Bhil. Hence, attending Gram Sabha meetings is of little use. Further, domination by a particular group dissuades others from participating fully in the deliberations.

- The Sarpanch and his cronies often use existing rules and provisions of quorum to their advantage and use Gram Sabha meetings to ratify decisions that suit them.

The Sarpanch has emerged as the key person: he alone prepares the list of beneficiaries for all the development schemes and also implements them. It is a common impression in Jhabua district that the Sarpanch has made a lot of money for himself out of these schemes, and as a result, the ordinary Bhil has got very little by way of his entitlements from the government.

Information in the Bhil villages is controlled by a handful of people—the Sarpanch and the government functionaries. Since the information is controlled, the benefits of the schemes are, almost always, cornered by these persons who have the information about them. The government functionaries, in any case, are keen on making money for themselves out of the government schemes. The Sarpanch is interested in favouring his cronies with whom he can share the largesse. The unhappy result is that the benefits of the government schemes do not reach the Bhils who are entitled to them.

This is a pity because Jhabua is a district where all the developmental schemes of the Government of Madhya Pradesh and the Government of India are being implemented. A lot of funds are released for the implementation of these schemes, but very little reaches the Bhil, the main reason being that the average Bhil is uninformed about what the schemes offer and how to avail of the benefits. Also the Gram Sabha, in which all the Bhil of the village are supposed to come together to decide who gets what, has ceased to be the democratic platform it is designed to be.

Scepticism of the Bhil towards the schemes worsens the situation. Having gone through the vicious cycle of exploitation at the hands of the Sarpanch and the government functionaries, the Bhil has lost faith in them, and they prefer to deal with the moneylender rather than the bank, the *Badwas* rather than the doctors, and the middleman rather than the development functionary of the government. As a result, they continue to remain vulnerable and poor. According to the Human Development Report of the Government of Madhya Pradesh, the human development index of Jhabua district is the lowest for the entire state (see Fig. 3.13).

Fig. 3.13 Human Development Index of Jhabua District

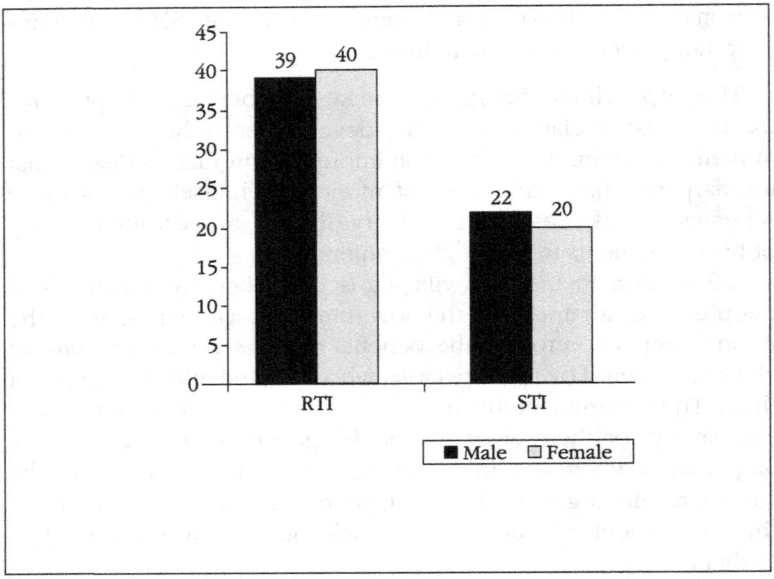

Source: Government of Madhya Pradesh (2002)

Livelihood

That brings us to the larger question of the livelihood of the Bhil. People need sustainable livelihood to ensure that they are able to access basic resources to ensure a decent life that is free of disease, hunger, squalor, poverty, deprivation and denial of basic human rights. Livelihood would, in that sense, be the sum total of both the employment portfolio of a household and what it earns in monetary terms, and what it receives as entitlements from the government as citizens of the state.

What about the livelihood of the Bhil? They have a large population dependent on agriculture, and therefore, its quality and productivity come to play a significant role. On both these counts, Jhabua scores very badly. The irrigation intensity is unacceptably low. The number of small holdings and the area under them has been increasing. Forest-based livelihoods have dwindled. Non-farm employment is negligible with hardly any organised employment. The entitlements from the government are so porous and flimsy that they do not contribute in any fair measure to the sustainability of the Bhil livelihoods. Overall, the picture is dismal.

Meanwhile, the issue of livelihood is further intensified by the increasing population. The Census of 2001 puts the population of Jhabua district at just under 14 lakh, showing a growth rate of about 24 per cent over the decade. This places the annual increase of population at 2.2 per cent or 28,000 persons per annum. Assuming that the level of main and marginal workers in the workforce remains the same, there would be, on an average, an addition of at least 20,000 to the workforce in Jhabua every year. This is a huge burden for the already precarious livelihood structure of Jhabua district.

As if to prove how precarious the Bhil's livelihood is, drought has been a regular visitor to the district in 1966, 1972, 1973, 1985, 1986, 1987, 1988, 1994, 1995, 1999 and 2000. During the drought periods, rainfall is scanty and intermittent. As a result, the *kharif* crop gets destroyed; it means the loss of the maize crop—the staple diet of the Bhil. It also means that there is acute drinking water shortage, and not enough water and fodder for the cattle. On the whole, the drought means considerable loss of water and fodder, loss of animal lives and untold misery for the Bhil.

There is Bhil folklore surrounding the drought spells in the village. According to the legend, the ancestors of the Bhil fought with the god of the clouds, and scanty rains with every passing year are god's way of punishing them. It is obviously apocryphal, but for a story that is steeped in Bhil legend, it is subtly nuanced. The story, told so simply, actually delineates an intense issue of intergenerational equity. The Bhil's ancestors indulged in acts of decimating vegetation and depleting water: the very sources that give the Bhil his livelihood. And for his ancestor's indulgences, the Bhil is condemned, as if by divine imprimatur, to pay penalties year after year.

Part-II

4

Empowerment—A Conceptual Framework

In the last three chapters, we have discussed about the varying dimensions of disempowerment of the Bhils, including their lack of food, education and health. The deprivations that the Bhil are subjected to severely limit what Amartya Sen calls the 'capabilities that a person has, that is, the substantive freedoms he or she enjoys to lead the kind of life he or she values' (Sen 1985: 187). Empowered people, on the other hand, have freedom of choice and action which enables them to influence the course of their lives and livelihoods, and also the key decisions which affect them. That is why empowerment of the Bhil is of critical importance.

A conceptual framework

This chapter attempts to define empowerment, identify elements that have, individually or in synergy, empowered people in successful initiatives. Based on an analysis of these elements, it provides a conceptual framework for empowerment. It also points out that although the conditions vary for empowerment in different contexts, its major influencing conditions can be identified. Finally, it looks at its vulnerability and suggests how efforts can be made to overcome it as part of an empowerment strategy.

What is empowerment?

The term 'empowerment' is a construct which has currency in various academic disciplines such as psychology, economics, studies of social movements and organisations; routinely used by politicians, presidents, poets, psychologists, civil society activists and development practitioners, it is also applied in practical disciplines like community development. A review of literature across both scholarly and practical disciplines does not, however, yield a clear definition of the concept along disciplinary lines (See Page and Czuba 1999: 1). What is even more confusing is that understanding of empowerment varies greatly among these perspectives; very often, the meaning of the term is assumed rather than defined. It is important, however, to understand the concept, at least broadly, so as to have some clarity on what constitutes it.

Understanding power

Power is the key word in the term 'empowerment'. As Dodd and Gutierez say, 'Although empowerment has been a social work buzzword"since the 1960s, most members of the profession have not taken serious, systematic look at the major root of the word power", yet, understanding power is essential before one can discuss what empowering practice is' (Dodd and Gutierez 1990: 121). In traditional social sciences, power is defined as: *(a)* the ability to get what one wants; and *(b)* the ability to influence others to think, feel, act and/or believe in ways that further one's interests (Parenti 1978: 21). This is in a relational context and reflected in three kinds of power—personal, social and political.

Personal power is based on the concept of perceived self-efficacy (Bandura 1981: 12). Individuals, who are personally powerful, know how to get what they want and how to influence others in ways that further their own interests. On the other hand, individuals who are personally powerless tend to gravitate to activities that are challenging but give them up when confronted with difficulty. This is either because they cannot get what they want or they are not in a position to influence others in ways that could further their own interests.

Social power stems from social resources that individuals bring to bear on relationships. There are five bases of social power—reward, coercive, legitimate, referent and expel power. They derive respectively, from: *(a)* the ability to provide rewards or avoid punishments; *(b)* the ability to impose punishments; *(c)* the normative; role-oriented values; *(d)* a degree of conformity in identifying with others; and *(e)* a degree of credibility or informational influence (French and Raven 1959: 87). One way of looking at these bases of social power—rather fashionable these days—is in terms of an exchange in which each individual in a relationship gives rewards to the other.

In that context, an individual who bags more rewards wields a power advantage in the relationship. However, there is a risk in viewing social power in this manner. To think that all relationships stem from an extremely asymmetrical power relationship begs the inference that the weaker individual stands to gain absolutely nothing (Parenti 1978: 27). In that case, there is no exchange involved, and worse still, the risk for the disempowered is almost overwhelming.

Political power is essentially about who determines who gets what or influences whom. Viewed from this perspective, power belongs only to those who make policy and set standards. Those who are politically powerful are able to maintain and perpetuate the status quo, which they, then, hold up as the only acceptable system and one which the powerless have no choice but to accept (Sahay 1998: 19–20).

It is important that we understand some of the important implications that flow from such interpretations of personal, social and political power. First, they imply that power is inherent in positions or people. Second, that power cannot change. Third, power is viewed as a zero-sum; this conception means that it will continue to remain in the hands of the powerful unless they give it up. What are the consequences if we interpret power in this manner? There are mainly two. First, if power cannot change, or if it is inherent in positions or people, then there is no possibility of empowerment. In other words, empowerment can take place only if power can change. Second, if power cannot expand, empowerment is not possible. The concept of empowerment would depend on the assumption that power can expand (Page and Czuba 1999: 2).

To have a clearer understanding of power, we need to go back to what Max Weber had said. According to him, power is related to our ability to make others do what we want (Weber 1946: 18). Building on this, traditional social science emphasised power as influence and control; as a result, power was treated as a commodity or structure divorced from human action (Lips 1991: 22). If power is conceived thus, it becomes unchanging or unchangeable. Weber, however, gives us a key formulation that helps us to go beyond this conundrum. He recognises that power exists within the context of a relationship between people or things. What this means is that power does not exist in isolation, nor is it inherent in individuals. Therefore, power and power relationships can change because it is created in relationships. In that case, empowerment as a process of change becomes a meaningful concept (Page and Czuba 1999: 2).

As indicated earlier, the concept of empowerment would also depend upon whether power can expand. Regarding power as zero-sum would imply that most people would not be in a position to experience power. Admittedly, this is one of the ways in which power is sometimes experienced, but the problem with such a formulation is that it does not encapsulate ways in which it is experienced in most interactions. Contemporary research on power has brought forth new perspectives that reflect aspects of power that are not zero-sum, but shared. Feminists, grassroots organisations and social and ethnic groups emphasise aspects of power that are characterised by collaboration, sharing and mutuality. What these aspects of power reflect is that gaining power actually strengthens the power of others rather than diminishing it. It is this aspect which is experienced in relationships that gives us the possibility of empowerment (Page and Czuba 1999: 2).

Defining empowerment

A review of definitions of empowerment reveals a great deal of diversity. Some definitions focus on the relational level, stressing the ability to negotiate and influence relationships and decisions. Some others explore empowerment at the personal level, involving a sense of self-confidence and capacity. Most definitions take into account structural inequalities that affect entire social groups rather than focus on individual characteristics. Some definitions describe dimensions that define the capacity to exercise strategic life choices such as access to resources, agency and outcomes. This diversity stems from the fact that the term 'empowerment' has different meanings in different socio-cultural and political contexts (Narayan 2002: 13–14). Empowerment is an intrinsic value; it also has instrumental value. It is relevant at the level of individual; it is also relevant at the collective level. Empowerment can be economic, social or political. There are important gender differences in the causes, forms and consequences of empowerment and disempowerment. It is not surprising, therefore, that there are so many definitions of empowerment.

In the broadest sense, empowerment is the expansion of freedom of choice and action. It means enhancing one's authority and control over resources and decisions that affect one's life. When people begin to exercise real choice, they acquire increased control over their lives. People who are disempowered do not have choices because they lack assets and capabilities and do not have the power to negotiate better terms for themselves *vis-à-vis* institutions that affect their lives. Since such powerlessness is embedded in the nature of their institutional relations, an institutional definition seems to be the most appropriate. So, empowerment can be defined as an expansion of assets and capabilities of the poor to participate in, negotiate with, influence, control, and hold accountable institutions that affect their lives (Narayan 2002: 14).

Assets and capabilities

Acquisition of assets and capabilities is necessary to increase well-being, security and self-confidence so that the poor can negotiate with those more powerful. Assets and capabilities are generally of the following kinds:

(a) Natural assets such as land;
(b) Physical assets such as infrastructure;
(c) Financial assets such as savings and access to credit;
(d) Human capabilities such as the capacity for basic labour, skills and good health;
(e) Social capabilities such as social belonging, leadership, relations of trust, a sense of identity, values that give meaning to life and the capacity to organise;
(f) Political capability such as the capacity to represent oneself or others, have access to information, form associations and participate in political life (World Development Report 2000: 14).

Assets like land, housing, livestock and savings enable people to cope with disasters and expand the horizon of their choices (Narayan 2002: 14). Lack of assets is both a cause and an outcome of disempowerment. The poor lack assets because of inequalities in the distribution of wealth and benefits of public action. Low assets and capabilities and low income are mutually reinforcing: low education leads to low income, which leads to poor health and reduced educational opportunities. Bad health, deficiency in skills and inadequate access to basic services reflect deprivations that keep them from leading the kind of life that most people take for granted.

A dearth of assets and capabilities results in a lack of voice, power and independence. This, in turn, makes the poor susceptible to humiliation, shabby treatment and exploitation by the state and society. As they are not in a position to take advantage of new economic opportunities, they find it difficult to participate in public affairs and to articulate their interests; as a result, their interests go disregarded. Social norms and barriers also contribute to voicelessness and powerlessness (Narayan 2002: 14).

As if to compound matters, vulnerability is the constant companion of material deprivation. The poor cultivate lands of marginal productivity in adverse conditions of scanty and uncertain rainfall. As a result, condition of their livelihood is at best precarious; this coupled with their inability to cope with shocks or mitigate risk stems from and reinforces the causes of other dimensions of their disempowerment. Vulnerability creates a vicious circle in which short-term actions to cope with it add to the deprivation of the poor people in the long-term. For example, at the time of crisis, the poor take their children out of school so as to get some extra income. They carry out distress sales of their land or livestock. The food intake of these families goes down to levels that are detrimental to health (The World Bank: 146).

Assets and capabilities can also be collective. For the poor, in particular, the capacity to work together in order to solve problems is a critical collective capability that they need badly in view of their meagre resources and marginalised position in society. It is in this context that social capital—the norms and networks that make possible collective action, which allows the poor and the marginalised to increase their access to resources and opportunities and participate in the process of governance—becomes important. One has to distinguish between different dimensions of social capital within and between communities. The strong ties connecting family members, neighbours and close friends are called *bonding social capital:* that which connects people who share similar demographic characteristics. The weak ties connecting individuals from different ethnic and occupational backgrounds are referred to as bridging social capital. This implies horizontal connections to people with broadly comparable economic status and political power. A third dimension— linking social capital—consists of the vertical ties between poor people and people in influential positions in formal organisations such as government departments and financial institutions (The World Bank 2000: 128). Linking social capital captures an important feature of the functioning of poor and marginalised communities in which the members are usually excluded from the decision-making process that affects their lives.

Poor people are good at bonding social capital (Narayan 2002: 15). They do this by establishing close ties with others who share the same characteristics as themselves. Such bonding helps them to cope with their disempowerment. Sometimes, the groups to which the poor belong bridge social capital by establishing ties with groups unlike themselves, but these ties are often unequal, resulting in patron–client relationships. When groups of poor people link with organisations of the state, civil society or the private sector, they are in a position to mobilise additional resources, and are able to participate in the societal processes (Narayan 2002: 15).

Institutional reform and empowerment

The institutions that affect the lives of the deprived are both formal and informal. The formal institutions consist of the laws and rules of the state, private sector and civil society organisations. The informal institutions include norms of inferior or superior status, expectations of bribes, kinship networks and informal restrictions placed on the basis of gender. The policies of the state and the culture of the institutions of the state determine, to a very large extent, the behaviour of the other institutional actors like the private sector and civil society organisations. When the policies of the state cater to the interests of the rich and powerful, and when the culture of the institutions of the state is characterised by corruption, exclusion and discrimination, its policies and programmes do not lead to the empowerment of the poor. It is in this context that it becomes important to address the culture, values and ethics of the institutions, since these can always subvert formal rules (Narayan 2002: 16). What is needed for the empowerment of the poor are institutions that are sensitive to their interests.

Poor people are generally excluded from participating in the institutions of the state which take decisions that affect the lives and livelihoods of the poor. For example, the community development programmes of the 1950s and 1960s regarded the poor as beneficiaries, but not as participants in the development process. The community development programmes were planned and executed in a top-down mode of decision-making that made no allowance for the local conditions and paid no heed to the specific needs of the local community. Due to the traditional power structures in the rural areas and the local barriers, it was commonplace for the bulk of the inputs to the villages under the community development programmes to be absorbed by the better-off sections of the rural communities (UNESCAP 2006: 3).

Development strategies in the 1970s and 1980s such as the integrated rural development programmes, the basic needs programmes and rural infrastructure projects did make an attempt to use local resources, but these policies and programmes continued to treat the poor as 'objects' or 'target groups' to whom development was sought to be delivered by outsiders. These policies and programmes viewed the rural poor as passive recipients who needed outsiders to come to their assistance; predictably enough, the approach did not empower the rural poor (UNESCAP 2006: 3). What was needed, instead, was a set of systemic reforms that would have enabled the rural poor and socially marginalised to participate in the state institutions and articulate their interests and aspirations.

Intermediate civil society organisations have critical roles to play in supporting and enhancing the assets and capabilities of the deprived, translating and interpreting information to them, and helping in linking them to the institutions of the state. In the implementation of most anti-poverty programmes, civil society organisations such as NGOs have become central actors. In the 1980s and 1990s, NGOs worked to develop awareness and motivation among the rural poor in order to empower them. As a result, the poor households and associated rural communities were sensitised to their dignity and rights as bonafide members of society in areas where these organisations worked (UNESCAP 2006: 3). These civil society intermediaries were thus instrumental in forging a link that was conspicious by its absence in rural development schemes for empowering the poor.

The association of civil society organisations in developmental efforts assumes importance as the majority of the rural poor are disorganised and have little or no training. They are, therefore, unable to take advantage of the range of opportunities and resources that are available from the government or other sources. This is made even more difficult by complicated rules and regulations that permeate the administration of developmental schemes. Civil society organisations have been helpful in this respect by working in areas of social mobilisation, awareness raising and income generation (UNESCAP 2006: 3). Several types of rural credit, inputs, extension and services are available at the local level through mobilisation by civil society organisations. Self-Help Groups have succeeded in mobilising a considerable amount of rural savings through their regular savings scheme. A large number of micro-finance agencies are also working at the local level to provide small credit to the rural poor.

On the whole, these intermediate civil society organisations have demonstrated techniques for reaching out to the rural poor and cultivating in them a sense of self-reliance, self-confidence and awareness. In addition, by providing cost-effective social services and by engineering basic economic and social reforms, they have helped in improving the economic and social status of the rural poor. However, these civil society organisations have to stay vigilant to ensure that they really reflect the interests of the rural poor in their activities (The World Bank 2000: 111). It is equally important that they are accountable to the poor that they seek to represent.

Several lessons emerge from the working of these intermediate civil society organisations in the context of institutional reform and empowerment. The needy (instead of poor people) need to be organised into viable groups. Community participation, self-reliance and self-help are *sin qua non* for empowerment. Women, especially, should be made bankable and given the broad responsibilities for community development. As most NGO initiatives have succeeded in empowering the poor, even government organisations have now started implementing activities in conjunction with them (UNESCAP 2006: 4). Much depends, however, on whether these civil society intermediaries have a vision. Only then can they empower the poor during the project implementation period and enable them to sustain the process even after the completion of the development project.

In the institutional context, social and cultural aspects are particularly important for empowerment exercises (Narayan 2002: 17). While building assets and capabilities, it must be ensured that institutional structures do reflect local norms, values and behaviour; in particular, they should build on the cultural strengths of the rural poor. For example, the skills sought to be inculcated in awareness-raising and training should be imparted through participatory methods, in which local knowledge mainly based on experience needs to be gainfully combined with outside knowledge. It is necessary that the empowering approach should put the poor at the centre of development and view them as the most important resource rather than as mere passive recipients. The approach should be to build on their knowledge, skills, values, initiatives, and motivation in order to solve problems and manage resources. It should treat the poor as worthy of honour, respect and dignity.

Since successful institutional strategies to empower the poor depend so critically on the social, institutional and cultural context of the local community, they vary from case to case (Narayan 2002: 17). In other words, there can be no single institutional model for empowerment. The challenge, then, is to identify the key elements of empowerment across social, institutional and cultural contexts and incorporate these elements while putting in place an institutional design for empowerment.

Elements of empowerment

Most successful initiatives for empowering the poor have four elements of empowerment. They are *(a)* access to information; *(b)* inclusion and participation; *(c)* accountability; and *(d)* local organisational capacity (Narayan 2002: 18). It is also important that these four elements act in synergy and not in isolation.

Access to information

Information that flows from the government to the people and from the people to the government is a key element of empowerment. Armed with information, the deprived will be in a better position to take advantage of opportunities, avail of services, exercise their rights, negotiate effectively, and hold the state authorities to account. In the absence of relevant information presented in forms that can be understood, it is impossible for the poor to take effective action. Dissemination of information should be through means such as group discussions, story-telling, poetry, debates and street theatre. Timely access to information in the local language at the local level is also important. Critical areas that need to be covered are: information about basic public services, financial services, likely markets for their produce and prevailing prices. Information and communication technologies can be used effectively to disseminate such information to poor people.

Inclusion and participation

Inclusion asks the question: Who is included? In an empowering strategy, the approach should be to regard the inclusion of poor people as critical in agenda setting, defining priorities and allocating resources at the local level so as to ensure that developmental expenditure is spent in accordance with local priorities and directed at a commitment to change. Participation addresses the question about how people are included and the role they play, if included. An empowering approach to participation should ensure that the poor are given authority and control over resources at the local level—particularly financial resources. This would enhance their capacity to engage in society's power structure and articulate their interests and aspirations.

Accountability

Accountability has three dimensions. First, public functionaries should be made accountable and responsive to users; corruption and harassment should be curbed; and the authority of the state should be used to redistribute resources for activities to benefit the poor. Second, there should be decentralised means for broad participation of the poor in the delivery of public services and mechanisms for eliminating the scope for the capture of the decision-making process by the local elite. Third, conditions should be created to encourage the participation of the poor in the process of governance, and articulation of their interests. There are three main mechanisms to ensure accountability: political, administrative and public. Political accountability is through elections. Administrative accountability is through internal accountability mechanisms, both horizontal and vertical, within and between agencies. Public accountability mechanisms hold the government apparatus accountable to the people. The empowering strategy should be such that these three accountability mechanisms are present and work in tandem to empower the poor.

Local organisational capacity

Local organisational capacity denotes the ability of the people to work together, organise themselves, and mobilise resources to solve problems of collective interest. Since the poor are very often outside the formal system, they turn to each other for support and strength to solve their problems. It so happens that whenever poor people organise themselves, they do it informally. Organisations of poor people are good at meeting the survival needs of the poor, but they have very little resources and virtually no technical strength. In addition, they often lack the capacity of bridging and linking socially, with the result that they may not be in a position to influence government decision-making and acquire collective bargaining power. If these constraints are taken care of, local organisational capacity is the key to empowerment, and organisations of the poor become important players in the institutional context.

Empowerment framework

As discussed, institutional reform to support empowerment would involve making radical changes in the relationship between the

Fig. 4.1 Empowerment Framework

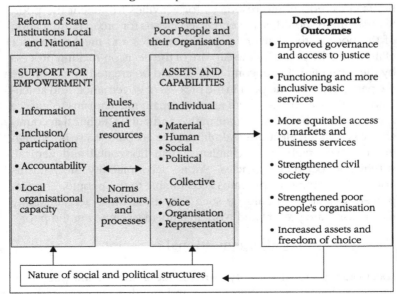

Source: Narayan (2002)

state and the poor and their organisations. The change should be transformative enough to focus on investing in poor people's assets and capabilities—both individual and collective—so as to enable them to participate effectively in society's power structures and interact with the government in order to strengthen the demand side of governance (Narayan 2002: 23). See Fig. 4.1 to see an empowerment framework that sets out the relationship between institutions, empowerment and improved development outcomes for the poor.

Empowerment in context: Conditions

As indicated, there is no single model for empowerment and no fixed blueprint for reforms. However, there are many examples of successful efforts to empower the deprived by increasing their freedom of choice and action in varying contexts. The variations notwithstanding, it is possible to identify major influencing conditions based on the experience of successful efforts (Narayan 2002: 24).

Motivation

The poor need to be encouraged and motivated to form groups, which should be the principal institutional mechanism for empowerment. Poor people would then begin to understand the important role of groups and the routine interactions they have in these groups, and the power that accrues to them through their assertion and articulation as productive members of the society (UNESCAP 2006: 4).

Self-management

The groups should be capable of electing their own leaders and managing their own activities like conducting meetings, record-keeping, savings and credit operations, conflict resolution and management of community property resources (UNESCAP 2006: 4). In the initial stages, civil society intermediaries could assist them in developing a system. But once the learning curve is over, the groups should be in a position to manage things on their own and exercise full autonomy over the system.

Resource mobilisation

Resources for funding the activities of the group need to be mobilised through regular savings of the members and voluntary contributions. The process also creates social capital. The pooling of resources of the group makes the members realise that each individual member owns resources, which, when combined, enhance the individual position of each member enormously. What is needed is to evolve an appropriate system of resource generation, allocation and utilisation so that all the members have equal opportunities (UNESCAP 2006: 4). This has the merit of ensuring ownership and sustainability.

Support

Groups of poor people require support in several ways. As experience has it, major impediments to the successful functioning of such groups are lack of time, resources, information, and absence of suitable support. What the groups need is technical assistance and skill building in order to become sustainable and effective. They need help in scaling up their membership, range of functions, awareness-raising and training. Such support is often provided by civil society intermediaries. It is helpful that the social and educational background of people working in civil society organisations enables them to interact seamlessly with the institutions of state and society, which in turn assists in creating bridges between the poor people's groups, institutions of state and society and outside agencies (The World Bank 2000: 110).

Decentralisation

Decentralisation moves decision-making closer to people, and in the process, public decisions reflect local priorities (Narayan 2002: 25). It also makes it possible for the members of the community to participate in local governance. Departments delivering public services are brought closer to the poor, thereby enhancing people's control of the services to which they have a right. But this calls for a strengthening of local capacity and devolution of financial resources. This also requires measures to avoid the capture of the apparatus by the local elite. For decentralisation to be successful, it needs to be combined with institutional mechanisms for people to participate in the process of governance.

It needs to be realised, however, that decentralisation by itself is not a goal of development but a mere means of enhancing the efficiency of the governmental apparatus. Decentralisation can make the institutions of the state more responsive to the poor, but only if they are in a position to hold government functionaries accountable and their participation in the process of development assured. In that sense, decentralisation can be a very powerful tool for achieving development goals in ways that cater to the needs of the local community by empowering the people who are best qualified to make decisions suitable to their own needs.

Vulnerability

We have so far discussed empowerment in terms of building assets and capabilities as well as enhancing the capacity of the poor to influence the institutions of state and society by strengthening their participation in processes of local governance and decision-making. The poor people are also vulnerable to external and largely uncontrollable events—bad weather and natural disasters—which exacerbate their material deprivation and weakens their bargaining position (The World Bank 2000: 3). That is why reducing poor people's vulnerability to risks by institutionalising mechanisms to help them cope with it is important to their empowerment.

Dreading the future— but knowing that a crisis may strike and that one will not be able cope —is an integral part of the life of the poor (The World Bank 2000: 135). Their low income means they are not in a position to save and accumulate assets; this, in turn, detracts from the deprived's ability to deal with crisis when it strikes. They do not have the resources to handle a crisis. Condemned to the most marginal lands and crowded areas with bad infrastructure, the poor are most at risk. When a crisis occurs, they need immediate increments to their income, or in the alternative they are forced to reduce their expenditure. This being the case most often, the damage is irreversible in terms of their economic and human development prospects.

Thus, these situations lead to child labour and malnourishment with lasting damage to children and, often, disintegration of families. When a crisis strikes and the poor cannot borrow from moneylenders or when adult unemployment is high or wages low, children are pulled out of school and sent to work. The forfeited schooling leads to a lifelong loss in earning ability for these children. This perpetuates poverty and vulnerability across generations. It is sad but true that in poor households, women are more afflicted than men during times of such crises. The increase in prices of food during crises often results in reduced nutrition intake for women as compared to others in the family.

Natural disasters account for most of the suffering, on a short-term as well as on a long-term basis; they detract from the ability of the poor to extricate themselves out of the poverty trap in the long-term by depleting their human and physical assets. Integral to any empowerment approach should, thus, be initiatives to prevent, mitigate and manage natural disasters, and as long-term measures, to establish safety nets that enable the poor to cope when disasters strike.

Droughts

Drought is a regular visitor to the district and causes hardship to the poor Bhil; this is also true of most parts of the country. It involves a period of distress ranging from weeks to months and sometimes years and is caused by lack of rainfall. The number of rainy days forms a very important basis for assessing drought situations. For example, India receives, on an average, around 1,050 mm of rainfall, which is higher than what many arts of the world receive. But about 30 per cent of the country receives less than 750 mm of rainfall, 42 per cent receives between 750 and 1,250 mm, 20 per cent receives 1,200 mm and 2,000 mm and 8 per cent receives 2,000 mm or more of rainfall (NRSA 2002). The problem, therefore, is the distribution of the rainfall and its erratic nature, with the result that over half of the country is arid and semi-arid, resulting in frequent droughts or drought-like situations.

Watershed development

60 per cent of the country's net sown area—about 85 million hectares—is rainfed (Joshi 2006). As a production system, rainfed farming has performed very poorly on every count—in terms of factor use, factor productivity, capital formation, employment generation, penetration of technology, innovation, stability and

sustainability. As if to compound matters, such a production system is the only source of food and livelihood for millions of families in India's villages. It is the primary occupation for an overwhelming majority of Scheduled Tribes, and it is not surprising that about half the Scheduled Tribes population remains below the poverty line and is chronically food insecure. About 60 per cent of the families below the poverty line in the country are farmers and an overwhelming majority of the practitioners do rainfed farming (Joshi 2006).

Raising the productivity of rainfed lands has been and remains one of the most difficult challenges to agricultural development. Soil conservation and improved dry farming techniques have long been recognised as essential for the purpose. Efforts to propagate them, dating back to the 1930s, focused mainly on the adoption of these measures by individual farmers (Vaidyanathan 2006). This approach continued to be the main thrust of the programmes for rainfed agriculture in the main phases of planning, but to little effect. It came to be recognised that basic improvements in soil and moisture conservation have to be planned on an area basis. A variety of special programmes for the development of wastelands, drought-prone areas and desert areas were launched at different times. During the 1970s, the concept of integrated watershed development came into prominence after the Indian Council of Agricultural Research (ICAR) advocated it as an appropriate strategy and implemented model projects to demonstrate the application of this concept in various locations (Vaidyanathan 2006).

Watershed development is considered as the most appropriate basis for sustainable management of land and water resources, and therefore the best way to combat drought. The objectives of integrated watershed development include control of soil erosion and land degradation, reclamation and rehabilitation of waste or degraded lands, land use revisions consistent with land capability, optional management of croplands, grasslands and forests, and conservation and management of water resources. All these measures do lead to optimal biomass production.

The idea of an integrated watershed is based on the concept of sustainable development, and sustainable development, in turn, is closely linked to the carrying capacity of the ecosystem. An ecosystem's carrying capacity provides the physical limits to economic development, and is the maximum rate of resource consumption that can be sustained on a permanent basis without affecting bio-productivity and ecological integrity. What it means is that improvement in the quality of life is possible only when the pattern and levels of production–consumption/conservation activities are compatible with the capacity of the natural environment as well as social preferences. To that extent, the planning process should ideally involve an integration of social expectations and ecological capabilities by minimising future differentials between realised and desired supply/demand patterns, infrastructure/congestion scenario, resource availability/use patterns and assimilative capacity/residual capacities (NRSA 2002: 9).

Does watershed development do this? What watershed development does is to establish a functional region through topological relationships. It evaluates the biophysical linkages of upland and downstream activities, as they are linked with the watershed by the hydrological cycle. Besides examining the chain of environmental impacts that result from land use activities and unplanned disturbances, it also provides a framework for analysing the effects of human interactions with the environment; environmental impacts within the watershed are in a feedback loop with changes in the social system (NRSA 2002: 15–17). More concretely, it conserves rainfall and the available run-off water so that the groundwater table is increased. It covers a large part of the watershed area with vegetation through plantation and natural regeneration—the strategy is one of integration through programmes like forestry, soil conservation, rural and community development and farming systems. On the whole, by utilising the watershed's natural resources such as land, water and vegetation, it mitigates the adverse effects of recurring drought.

Although the idea of watershed found acceptance as part of official policy in the early 1970s, it did not make much of a difference to the way the programmes were designed and managed, or to their impact on the quality and productivity of rainfed lands. There were a large number of schemes with similar or overlapping scope, and they were planned and implemented by numerous agencies in a fragmented and piecemeal manner (Vaidyanathan 2006: 2984). There is a general consensus that they did not make a significant impact in containing land degradation or increasing the productivity of rainfed lands. The communities which were meant to benefit were kept totally out of the process, and there was practically no monitoring or evaluation of physical works, their quality and impact.

The 'watershed' thinking of public policies and programmes is characterised by two flawed conceptual biases (Joshi 2006: 2987). One is the land bias, and the other the water bias. The land bias of the watershed thinking comes from the preoccupation with conventional soil conservation approach of safe disposal of run-off. And, more recently, watershed development has been equated with rainwater harvesting and conservation. The soil conservation bias comes from the long-standing concern of silting of dams, and about loss of topsoil, leading to expansion of wasteland and desertification. Conservation of water becomes the sole preoccupation of the water bias. While both these dimensions—husbandry of soil and conservation of water—are important, they should not be the sole objectives of watershed development. If that is the case, watershed development degenerates into husbanding of natural resources to optimise the elements; in other words, to extract the best from the land-based portion of the water cycle, albeit in a sustained manner. Normatively, the objective should be defined in a way to maximise present and future well-being of the largest number of people, especially the poor, for whom these resources represent the only means of livelihood.

In that sense, 'livelihood' should become the most important consideration in the watershed thinking. This should provide watershed development with a larger social perspective and purpose (Joshi 2006). By focusing on livelihoods, watershed should become the mainstay of poverty alleviation programmes, given the geographic and agro-ecological contours of poverty in India. That way, it will be the driver of decentralised growth with distribution. Livelihood, then, becomes the objective function that is optimised through watershed development, and it is only sustainability that should set the boundaries as a constraint. In that case, watershed development would not only enhance rainfed agriculture's contribution to economic growth but would also achieve that in an equitable manner which is also environmentally sustainable. It is only in such a context that watershed development deserves the centre stage for not only ecological restoration but also agriculture and rural development.

Clearly, watershed development needs to be reviewed in the context of a conceptual framework that organically links three important elements—bio-physical (natural resources management and ecological balance), socio-economic (productivity, agricultural growth and livelihood support), and institutional (sustainable resource use, equity and benefit cost sharing). The efficacy of any one of these elements would depend on how it is in synergy with the other elements. This only reinforces the interconnectedness of the biophysical, socio-economic and institutional aspects of watershed development within a larger conceptual and normative framework for natural resource-based sustainable development (Joy *et al.* 2006: 2996).

How does watershed development work as an empowering strategy? Interestingly, it has the potential of incorporating all the elements of an empowering strategy. By mitigating the adverse effects of recurring droughts, it would lessen the vulnerability of poor households, reduce and mitigate risks, and neutralise the impact of shocks. It would help in building the assets and capabilities of poor households, and enhance the capacity of the deprived to influence the institutions of state and society by strengthening their participation in local governance and decision-making.

But then, the question is: Do these watershed development projects lend themselves to people's participation? The answer is yes. They have tremendous potential for participation of stakeholders in the planning and implementation of the projects. This is possible only if the affected are made to realise what the stakes are. A comprehensive understanding of the threats in the watershed should bring about a commonality of purpose in working out solutions to meet them. In fact, planning for the watershed is the only way that would provide the affected an opportunity for getting involved in a participatory process for finding collective solutions. Such a participatory process, particularly one that proactively include the women and the poor, would create the right kind of environment in which they have the opportunity to build their assets and capabilities, engage in society's power structure, articulate their interests and empower themselves.

5

The Watershed Project

Jhabua district is drought prone. During droughts, the normal cropping area reduces up to 60 per cent; approximately 25 per cent of the villages face acute drinking water shortage. The problem stems from the inability of the terrain to conserve soil and water. Soil erosion too is a serious problem. Large-scale degradation of forest area has resulted in severe soil erosion, leaving behind 36 per cent of the arable land with practically no soil cover. All one sees in Jhabua is a succession of sandy and barren hill-slopes. They have no topsoil and therefore, no water-holding capacity. Although the district receives an annual rainfall of 600 to 800 mm, its erratic and unevenly distributed nature makes it drought prone. The watershed project, which was started in Jhabua in October 1994, was based on the assumption that by utilising the watershed's natural resources such as land, water and vegetation, the adverse effects of recurring drought would have been mitigated.

The model

The watershed project in Jhabua district followed a sustainable developmental model (see Fig. 5.1).

The vision

The vision of the watershed project was defined; it had four dimensions: environmental, economic, institutional and social.

- Environmental—the ecological balance in the watershed area is restored.
- Economic—improved and sustained livelihood status of the watershed community with special emphasis on the poor and women.
- Institutional—sustainable community organisation established.
- Social—an empowered community.

Fig 5.1 Sustainable Developmental Model

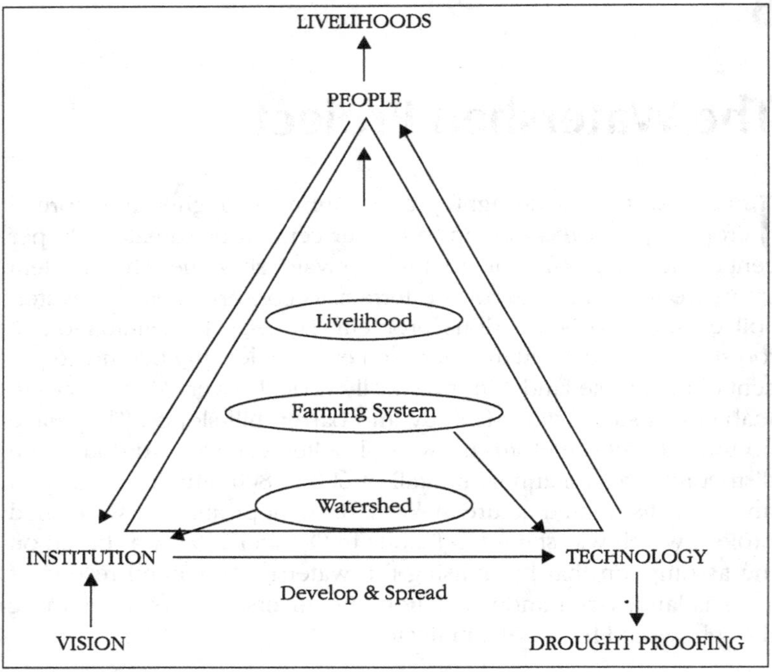

Source: Collector, Jhabua (1994)

The outcome

The outcome of this project was related to the vision statement. They were

- Environmental—optimum utilisation of the watershed's natural resources like land, water and vegetation that would mitigate the adverse effects of drought and prevent further ecological degradation leading towards restoration of the ecological balance.
- Economic—increased agriculture productivity.
- Institutional—sustained community action for the operation and maintenance of assets created and further development of the potential of the natural resources in the watershed.
- Social—increased level of awareness on social issues, and women and poor to be given due share in the development process of society.

The structure

In the project, a micro watershed is the unit for implementation: an area of about 500 hectares draining to a common point. A milli watershed is an aggregation of 10 adjacent micro watersheds with an area of 5,000 to 10,000 hectares. A macro watershed is 10 adjacent milli watersheds draining to a common point. For purposes of implementation, a Programme Implementing Agency (PIA) is in charge of each milli watershed.

Self-Help Groups and user groups were organised in each micro watershed. Watershed Associations were formed at the level of the micro watershed to ensure people's participation. It is the equivalent of the Gram Sabha for the watershed community; an organisation of all the people in the village or villages who had a stake in the watershed project. The Association appointed a Watershed Development Committee of 10 to 12 members with four to five members from the user groups, three to four members from Self-Help Groups, two to three members from the Gram Panchayat and one member from the PIA. The members of the Watershed Development Committee elected a chairman. An educated youth was appointed as the secretary of the committee as a paid employee. In every micro watershed, a volunteer was also appointed.

At the level of the milli watershed, the PIA is the core agency. A multi-dimensional task force works with the PIA. The task force has members from the various departments of the government such as forest, agriculture, fisheries, dairy, animal husbandry, soil conservation and tribal welfare. Full-time staff is also inducted from the District Rural Development Agency.

At the district level, the collector of Jhabua district is the mission leader of the project. The chief executive officer of the Zilla Parishad (ZP) is the nodal officer and project director for the project. The additional CEO, ZP is the district coordinator. The District Watershed Advisory Committee is the highest body for the project; it monitors the progress in the implementation of the watershed works. The president of the ZP is the patron of the Committee, and all the MLAs, MPs, the collector, the PIAs and the district-level officials of the development departments are members of the District Watershed Advisory Committee. The Committee which meets every fortnight reviews the progress of the project. There is also a District Watershed Technical Advisory Committee under the chairmanship of the CEO,

ZP, that looks into the technical problems encountered during the implementation of the project. The Committee meets on the last working day of every month and discusses the progress of the project with the PIAs (see Fig. 5.2).

Fig 5.2 Organisational Structure of the Project

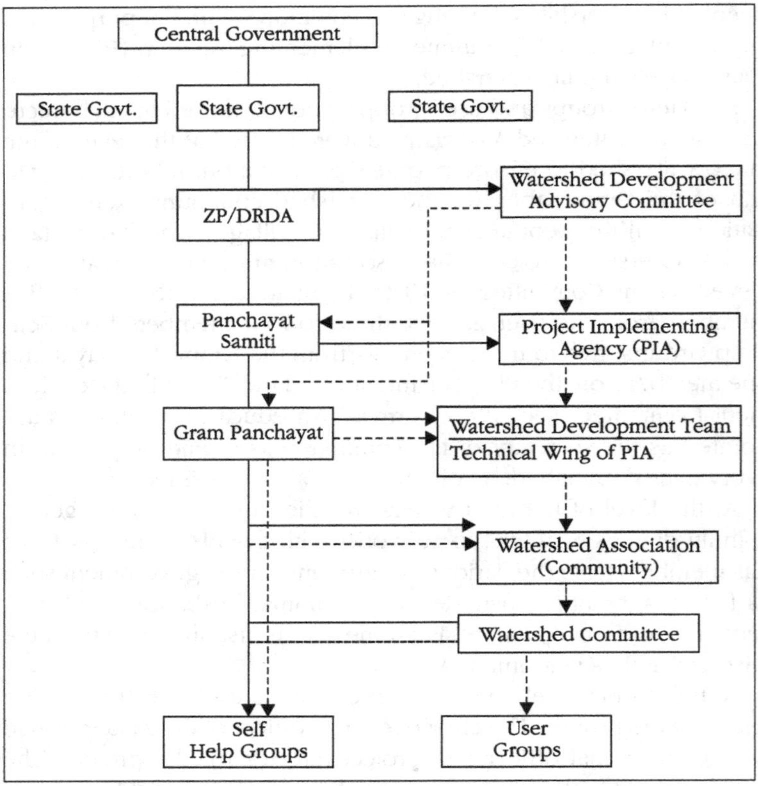

Source: Mahajan (1996)

Project contents

The stated goal of this project was optimum utilisation of the watershed's natural resources such as land, water and vegetation. It generated action plans with a view to harness the productive potential of the local land and water resources to sustainable levels. The plan consisted of a combination of water resources action plans and land resource action plans. Water resources' action plans addressed water needs through storage mechanisms and silt retention; they aimed at developing, conserving and efficiently utilising the available water on the surface, in the soil profile and the groundwater. The land resources action plans had several approaches. In case of sub-optional land use, the approach was to recommend upgradation, such as conversion of a single crop area. The other approach was to use vacant or whole lands for forest plantation, etc. The plans also recommended alternate land use plans such as introduction of improved tools, and contemporary crop and animal husbandry techniques. A common approach was to suggest the introduction of new agricultural practices but subject to local acceptance.

By way of water resources development, this project suggested small structures like farm ponds, minor irrigation tanks, gully plugs, check dams and other small water storing structures to address local needs for providing supplementary water for irrigation and drinking purposes. It emphasised that while the use of surface water would be the primary basis in the watershed development strategy, the groundwater would be conserved for emergencies only. The emphasis in the action plans was to raise the water table considerably, and develop the water resources to an extent that a sizable *kharif* area receives life saving irrigation in a long dry spell. The idea also was to ensure residual moisture for the *rabi* area under cultivation and enable a part of the *rabi* area to do an irrigated crop in a drought year.

In respect of land resources development, this project proposed ridge to valley treatment, and the total land area of the watershed to be treated under various soil conservation measures. It also proposed that the bulk of the common property resources should be taken over for maintenance, and a large portion of the watershed area should be covered under vegetation through plantation and natural regeneration. For this purpose, it suggested a number of pasture development measures and *silvi-pastoral* activities. The idea was to enable all the families in the watershed to meet their fuel and fodder requirements even under drought conditions.

On the whole, this project's goal was to mitigate the adverse effects of drought, prevent further ecological degradation and restore the ecological balance. In essence, it set forth a drought-proofing strategy, which incorporated dug wells as a recharging and irrigation tool as well as a low-cost watershed programme that converted surface to sub-surface water and then exploit it when required.

Biomass production

One of the major objectives of this project was to significantly increase the biomass production in the watershed area. Towards that end, it planned to ensure that: *(a)* at least 50 per cent of the total rainfall and 80–90 per cent of the available run-off in the watershed is conserved; *(b)* around 80 per cent of the total land area of the watershed is created under various soil conservation measures; *(c)* around 80 per cent of the common property resources (revenue/forest/panchayat land) are taken over for operation and maintenance; and *(d)* at least one-third of the watershed area is covered under vegetation through plantation and natural regeneration.

Integrated livelihood strategy

This project suggested the adoption of an integrated livelihood strategy for achieving improved and sustained livelihood status for the watershed community with special emphasis on employment being created and nurtured for the poor and the women. Apart from suggesting ways and means for doubling *kharif* and *rabi* production, the project recommended supplementing farming systems with allied activities such as dairying, fisheries, poultry, horticulture and goatry. It also proposed that all poor households be covered under the activities of Self-Help Groups and linked to livelihood means through income-generating schemes.

Increased farm productivity

Steps were put in by the project to increase farm productivity, agriculture and allied activities. Some of them were: *(a)* increase in the double cropping area, particularly in the case of land belonging to families below the poverty line; *(b) Rabi* irrigation potential to be increased; *(c)* survival irrigation for *kharif* crop to be boosted;

(d) increase in cropping intensity; (e) change in cropping pattern from crop to commercial; (f) adoption of appropriate land management practices as per the gradient and other attributes of the land; (g) increase in area under silvi-horticulture; (h) expansion in area under agro-forestry/horticulture; (i) adoption of livestock breed improvement programme for a large number of households; (j) motivating a substantial number of households to engage in off-farm activities like brick-making, bamboo collection of Non-Timber Forest Produce (NTFP) etc; (k) reduced dependency on money-lenders for productive and consumption loans; (l) generation of substantial employment locally as a result of watershed development and augmentation of natural resources; (m) one-third of the total household income generated through on-farm (poultry, dairy, goatry) and off-farm (brick/basket/leaf plate making, etc) activities as supplementary to the agricultural income; and (n) ensuring food security for all households of the watershed association in a drought year.

Increasing the level of awareness on social issues

Particular care was taken by the project to incorporate initiatives that would increase the level of awareness on social issues, achieve optional social goals and ensure that women and poor are given due share in the development process of the society. They were:

(a) Organising awareness camps on issues like education, health, hygiene, etc. at regular intervals;

(b) Ensuring regular visits and attendance of health workers, teachers, extension workers and the staff of the government departments;

(c) Increasing the rate of education among children, particularly girls;

(d) Minimising school drop-out rate;

(e) Increasing the awareness level about the importance of vaccination and family planning;

(f) Increasing awareness about health, education and other services, and ensuring high quality of public service delivery;

(g) Organising at least 80 per cent of women into Self-Help Groups;

(*h*) Ensuring women's active involvement in the planning, execution and monitoring of works;

(*i*) Creation of opportunities for women to air their views;

(*j*) Organising Prohibition campaigns at regular intervals;

(*k*) Reducing the level of liquor consumption;

(*l*) Reducing drudgery faced by women, i.e., fetching drinking water, fuel wood and fodder, to facilitate the participation of women in village development; and

(*m*) Decreasing the number of community conflicts and evolving mechanisms for resolving them internally without external interface.

Sustained community mobilisation and action

The watershed project stipulated that sustainable community institutions capable of identifying and addressing common development issues of the watershed community should be established and developed. It further ensured that these community institutions should be truly democratic in their functioning and should be sensitive to the causes of the poor and women. Listed below are the initiatives towards this end:

(*a*) All households in the watershed should be members of the Watershed Association.

(*b*) At least 80–90 per cent of the households, including the poor and women, should actively participate in the planning of the watershed.

(*c*) At least 80–90 per cent of the households, including the poor and women, should be given exposure to successful watersheds.

(*d*) At least 50 per cent of the households, including not less than 75 per cent of families below the poverty line, should receive skill training, both technical and managerial.

(*e*) At least 80 per cent of the households should be organised into Self-Help Groups and user groups.

(*f*) Self-Help Groups should generate substantial amount of savings and rotate credit among themselves.

(*g*) At least 70–80 per cent of the Self-Help Groups should mobilise external resources from NABARD or SJSY to meet the larger productive credit needs of the watershed community.

(h) All community organisations (watershed development committees, Self-Help Groups, user groups) should develop and use their own functional norms.

(i) Watershed Development Committees should actively participate in the implementation of the action plan.

(j) Watershed Development Committees should develop working modalities and get it approved by the Watershed Association.

(k) Watershed Development Committees should demonstrate their accountability to the Watershed Association by creating transparent systems of work approval, financial transactions, etc. (for example, the treatment plan approved by the Watershed Association may be painted on the wall of a common place like panchayat/school building. All financial transactions of a month may be read out in the monthly meeting of the Watershed Association).

(l) Watershed Development Committees should develop norms for the use of Gramkosh and distribution of benefits.

(m) User Groups and Watershed Development Committees should maintain all the structures created under the watershed project.

(n) Community organisations should ensure that the services of the line departments of the government converge at the watershed level.

(o) Marketing systems should be put in place for agricultural production to minimise the effect of middlemen.

(p) Enhance the trend in Watershed Development Committees, Self-Help Groups and User Groups from observer–discussant–decision-maker.

Assets and liabilities

This project also delineated a positive strategy to upgrade the assets and liabilities of the Bhil in the watershed area. Every Bhil has some land, and in the project area, steps were proposed to create necessary conditions for increased productivity in these lands. Plans were also mooted for reducing volatility in the productivity of these lands by proposing initiatives that would mitigate the adverse effects of drought and prevent further ecological degradation.

It also proposed steps for providing increased access to infra-structure for the Bhil in the watershed area. Regarding financial assets such as savings and access to credit, it proposed that the Self-Help Groups should generate substantial amount of savings and rotate it as credit among the members, in addition to mobilising external resources from NABARD and SJSY for meeting the larger productive credit needs.

Besides suggesting special initiatives for education, skill-formation and good health, it also proposed skill training, both technical and managerial, for at least 50 per cent of the households in the watershed area, including not less than 75 per cent of the families below the poverty line. Provisions were made for organising awareness camps on issues such as education, health and hygiene at regular intervals. This undertaking also proposed steps to increase the rate of education among children (particularly girls), and minimise the school drop out rate. There were steps to increase the awareness level about health, education and other services, besides ensuring the quality of public services.

As regards social capabilities such as social belonging, leadership, relations of trust, sense of identity, values that give meaning to life, and the capacity to organise, this project suggested sustained community mobilisation and participatory management of these community institutions. The project mooted that all households in the watershed should be members of these community institutions, and there should be an increasing trend in these institutions for the members to progress from observant to discussant to decision-maker. For acquiring political capability that includes the capacity to represent oneself or others, access information, form associations, and participate in the political life of a community, the project pro-posed steps: the strong community mobilisation and action that it advocated was to give the participants the necessary political cap-ability to empower themselves.

Participation in local governance and decision-making

This project proposed active participation of all households in the watershed community, including the women and poor in all the community institutions such as the Watershed Association, Watershed

Development Committees, Self-Help Groups and user groups and address its common institutional issues. It also proposed that most households should be organised into Self-Help Groups and user groups.

Vulnerability

Recurring droughts in Jhabua district, as discussed earlier, made the Bhil extremely vulnerable. The watershed project in Jhabua was, in fact, an attempt to mitigate the adverse effects of drought by optimally utilising the watershed's natural resources like land, water and vegetation. The undertaking also suggested concrete steps for drought-proofing such as increasing the ground water table, providing life-saving irrigation to the *kharif* crop in a long dry spell and ensuring that the *rabi* area requiring residual moisture (gram) is under cultivation and at least 50 per cent of the *rabi* area to do an irrigated crop in a drought year. Moreover, steps were also suggested for all families in the watershed to enable them to meet their fuel and fodder requirement from local sources even under drought conditions.

Training and exposure

An important feature of this project was the training of all those involved in the watershed activities. 5 per cent of the project outlay was earmarked for training of the participants and exposing them to successful watershed projects elsewhere in the country.

Distribution of project funds

The distribution of project funds, among various activities (see Table 5.1), ensured that 75 per cent of the project funds were spent

Table 5.1 Distribution of Project Funds

Entry Point Works	5%
Community Organisation	5%
Training	5%
Administration	10%
Watershed Works	75%

Source: Collector, Jhabua (1994)

by the Watershed Development Committees on the watershed works and 25 per cent by the Programme Implementing Agency (PIA).

When it came to financing this project, it mooted the idea of utilising the available funds to the district under the approved schemes of both the central and the state governments. The funds from schemes such as Employment Affirmation Scheme, Drought Prone Areas Programme, Integrated Wasteland Development Project, Development of Women and Children in Rural Areas (DWACRA) and Jawahar Rozgar Yojana were pooled and utilised to finance the project. The financing available in the Self-Help Group Scheme of NABARD was proposed to fund the credit activities of the Self-Help Groups.

Gramkosh and development account

The creation of funds from different sources was envisaged by this project. One being from the Gramkosh or the village fund, which consisted of contributions made by villagers from payments received as wages for watershed works or any other group activity in the project. The beneficiaries were required to contribute 50 per cent of their wage earnings from watershed work in their own land, and 25 per cent of their earnings from works in the common land.

Several other sources of financing of the Gramkosh were contemplated in this project that included:

- Amount received from the sale of grass grown on the common land of the village;
- Fine amounts received as a result of social control mechanisms; and
- Security amounts under the scheme of joint forest management.

This project laid down certain guidelines for the use of funds in the Gramkosh.

- Refinancing for Self-Help Groups;
- Payment to the guards of forest and common land;
- Social welfare activities;
- Honorarium to teachers of *Falia* schools;
- Establishment of grain bank;
- Loans for seeds, fertiliser and irrigation tools;
- Construction of community halls;

- Meeting drinking water problems; and
- Deepening of wells.

It merely suggested these as guidelines; it was left to the villagers to decide how the funds in the Gramkosh should be utilised.

The other fund mooted by it was the Vikas Khatha—the development account; a futuristic fund for the maintenance of the infrastructure created by the watershed project. It was designed as a recurring account from which money could be withdrawn only after the completion of the project. The beneficiaries were required to contribute 5 per cent of their wage earnings to this account. Thus, Vikas Khatha was an innovative feature, for in most government programmes no attention is paid to the maintenance of the assets after this project is completed.

The two funds—Gramkosh and Vikas Khatha—were seen in the project as being critical to the sustainability of the watershed project. They were designed to create conditions for the ownership of the watershed project by the beneficiaries; the means by which the beneficiaries could identify with the watershed activities. These two funds were important in yet another way; they were expected to influence the savings–investment cycle in the Bhil community. In the indebted Bhil setting, such savings had a special meaning, and they served a useful purpose: they were invested in activities of village development and reconstruction.

Self-help groups

In this project, the community base started with the Self-Help Groups—the thrift and credit groups—which were organised in each *falia*. By making the Self-Help Groups its base, it assumed that the savings and credit activity of the group- because of the multiple benefits it provided—would be the bond to join the members in a commonality of purpose.

The Self-Help Groups were expected to: inculcate the habit of saving among the Bhil women; build self-confidence in the women for becoming self-reliant; enable the group to give credit to the needy; give the Bhil women a sense of ownership of the watershed activities; participate in other group activities of the village; network with formal financial institutions to increase the capital stock of the group; and empower the Bhil women.

Some of the conditions the project stipulated for the successful working of the Self-Help Groups include:

- Efficient representation;
- Timely savings and loan repayment;
- A set of rules and definitive job chart for each member of the group;
- Maintenance of the records and registers such as the minutes of the meetings, records of loans given, record of repayment, record of deposits, and other registers to monitor and analyse progress;
- Homogeneous and local membership;
- Weekly meeting;
- Training;
- Transparent financial management;
- Intra-group coordination; and
- Simple and easily comprehensible functional norms and rules of the groups;

It also suggested certain guidelines about how the group should function:

- Not more than 15 members in a group;
- Members to deposit Rs 10 as membership fee and at least Rs 10 as monthly savings;
- Every member to be given a savings account passbook to record detailed information regarding savings, deposits, loan taken and loan repayment;
- The group to have a secretary and treasurer. They are to be held responsible for proper and regular upkeep of records;
- Savings account to be opened in any nationalised bank or post office or any branch of a cooperative bank. The account to be operated through three select members (two from the Self-Help Group and one from the watershed development committee);
- Money collected to be deposited either that very day or the next day;
- Only those who have deposited Rs 10 for 10 consecutive months are entitled to get a loan from the Self-Help Group;
- Repayment schedule to be decided by the group;
- Loanees to be decided through internal group dynamics. When the group decides on a loan, more than half of the members of the group should be present;

- A guarantee of at least two group members before sanctioning a loan;
- Loanee to make an application asking for the loan;
- Passbook to be kept in the office;
- Credit rates to be decided by the group;
- Action to be taken against defaulters; and
- Second loan to be given only when the first loan is completely repaid.

The PIAs were given a very special role in respect of the Self-Help Groups. They were expected to:

- Organise the groups, select the president, and design the framework of rules;
- Motivate the members;
- Give training to the members;
- Advise the group about the investment of the capital stock;
- Observe the working of the group, create awareness about its strength and help in correcting aberrations;
- Help the group in preparing an action plan; and
- Educate the members in the maintenance of accounts.

The resource base of the Self-Help Group was to consist of the savings of the members. But, since the group was expected to fund the credit needs of all its members, this project proposed augmenting the resource base of the Self-Help Groups in a number of ways. A revolving fund of Rs 1,00,000 was provided to each micro-watershed; the group could use the revolving fund on the basis of matching contributions. The group was also given Rs 25,000 under the DWACRA scheme. Assistance from DWACRA was available only after six months of the group coming into existence. The idea was that before getting DWACRA's assistance, the group should have started functioning.

The groups were linked to the credit facilities of the commercial banks under the NABARD Scheme of Self-Help Group Assistance. Lending to Self-Help Groups in the scheme is considered as main-stream credit operations under the priority sector. The rate of interest charged is 12 per cent, and the loan given by the commercial bank is in proportion to the savings that the group has mobilised; that is, the maximum ratio between the savings of the group and the loan from the bank is 1:4.

This project emphasised that the Self-Help Groups should mobilise the savings of their members to the maximum. This was so because acquiring 'personal savings' is a big boost to the Bhil woman who has, otherwise, no ownership rights over any kind of resource. Having money in the bank is something that instils a sense of economic self-confidence in the women. This, the project assumed, would enhance their status; also their capacity for independent earning was expected to play an important role in empowering them. It is apposite that the scheme is called *Baira ni Kuldi* in this project because in the Bhil language, *Baira* is the word for women while *Kuldi* means a kind of pot.

The role assigned to the Self-Help Groups in the project went much beyond their thrift and credit activities. In its design, they were seen not merely as thrift and credit groups, but as institutions of social reconstruction which were expected to outlast specific projects and ensure the development of the Bhil community in the future. In other words, the Self-Help Groups were regarded as the primary change agents; they were the life line of this project. The proper functioning of the Self-Help Groups, in that sense, was crucial to the success of the watershed project.

The organisation of the Bhil community into self-reliant institutions at the *falia* level was an important part of this project. Development of community-based organisations (CBO) was the first step; community awareness, organisation and mobilisation were the basic elements, which, in different proportions, constituted the change force in the design of this project.

The Self-Help Groups were the functional unit of all the CBOs; they formed the nucleus of all the activities, including planning, implementation and monitoring of this project at the level of the *falia*. While designing the project, care was taken to ensure that the existing kinship groups in the Bhil community were organised into Self-Help Groups. This was because traditional practices like *Halma* in the Bhil community, which are based on concept of shared labour and social exchange, are prevalent among Bhil kinship groups. Organising Self-Help Groups along the lines of existing kinship relationships was expected to facilitate the smooth functioning of the group; it was also to contribute to the sustainability of the institution.

Group formation was seen as a very important activity in this project. The assumption was that the required awareness would have been created in the process of group formation; it would give the Bhil the necessary understanding of their roles and responsibilities in the planning and implementation of this project. It was only after the groups were formed that discussion and action plans were prepared for the watershed activities.

Roles and responsibilities

This project spelt out the roles and responsibilities of the different players involved in its planning, implementation, monitoring and review. The government's role in it was seen as one of facilitation. It was expected to fund the project in the initial stages; allow the use of government property resources such as common land, forests, streams and rivers; provide technical assistance and review the progress of the project periodically in consultation with the elected representatives; find solutions to problems of coordination between agencies involved in the implementation of this project.

The role of the PIAs was to create awareness among the Bhil about this project; organise them into groups; increase their capacity through various interventions; act as a demand creator for the groups so that maximum benefits and resources both internal and external were generated optimally; and operate as a facilitator for it. However, the role assigned to the people was to organise themselves into groups; provide local leadership; and participate in its planning and implementation.

The role of the village community was to provide: local resources and facilities such as land, tools and other required materials; community's traditional knowledge on local conditions; physical labour—wage earning as well as contributory; its savings; and management of institutions such as the Self-Help Groups, user groups, grain banks and watershed development committees.

Logical framework analysis

One significant aspect of formulating the watershed project in Jhabua was the use of Logical Framework Analysis as a tool for its plan at the micro level. It covered the three areas of vision, outcome and activities. It set down the success indicators, the means of verification and the participatory methods that were to be applied for generating information to fix the objectives as well as to generate information for the base line. The analysis proposed was as in the following (see Table 5.2):

Table 5.2 Logical Framework Analysis

Logical Framework Analysis—Tool for Watershed Planning at the Micro Level

	Vision		
Vision (Post Project Scenario)	Success Indicators	Means of Verifications	Participatory Methods
Environmental			
• Ecological balance in the watershed is restored	• Adequate availability of drinking water for human beings and cattle in a drought year. • All families in the watershed able to meet fuel and fodder requirement from local sources even under drought conditions.	• Post-project evaluation study—comparison with baseline.	• Focused Group Discussion (FGD) • Resource mapping • Community Problem Analysis (CPA)
Economic			
• Improved and sustained livelihood status of the watershed community with special emphasis on the poor and women	• Significant improvement as per human development indicators • Food security ensured for all households of the watershed even in a drought year. • Households' income, especially of resource poor villagers (BPL), increased from X to Rs 2,000 p.m.	• Post-project Impact Assessment. • There could be: Self assessment by the PIA/by the Community and/or by External	• Focused Group Discussion (FGD) • Well-being ranking • Livelihood analysis of different S-E classes

Social			
· Empowered Community	• Cent per cent achievement of National Health Programs	• Post-project Impact Assessment	• Focused Group Discussion (FGD)
	• Cent per cent enrolment in schools with good quality of education		• Gender appraisal methods
	• Women and poor are empowered adequately to establish their role in the decision-making process of the village development		
	• New leadership emerged to provide direction to the village society as a whole		
	• High level of awareness about the socio-political issues		
	• Active stake in the Gram Panchayat		
Institutional			
• Sustainable community organisation established	• Institutionally sustainable community organisation capable of identifying and addressing common development issues	• Post-project Impact Assessment	• Focused Group Discussion (FGD)
	• Truly democratic (absence of patronage relationship)		• Chapati Relational Diagram
	• Sensitive to the causes of the poor and women		• Institutional history

Source: Collector, Jhabua (1994)

Table 5.3 Outcome Analysis

Outcome			
Outcome Results to be Achieved Out of the Activities	Success Indicators	Means of Verifications	Participatory Methods
Environmental •Optimum utilisation of the watershed's natural resources like land, water and vegetation, etc. that will mitigate the adverse effects of drought and prevent further ecological degradation leading towards restoration of ecological balance.	•Ground water table increased from X to Y. •At least 50% of the *kharif* area receives life-saving irrigation in a long dry spell •100% of the *rabi* area will be under cultivation requiring residual moisture (gram) and at least 50% of the *rabi* area can be under irrigated crop, if opted for, in a drought year. •All families in the watershed being able to meet fuel and fodder requirement from local sources even under drought conditions.	•Field survey/observation •Monitoring of wells •Field survey •Field survey	•Mapping of cropping area (*kharif, rabi*) •Transect •Seasonal analysis •Need analysis of fuel wood, fodder, etc. •Time line •Matrix ranking

Economic

• Increased farm (agriculture and allied) productivity	• Households' income, especially of resource poor villagers (BPL), increased from X to Y. • X% of the employment generated locally as a result of watershed development and augmentation of natural resources. • Irrigated area increases from X to Y. • Dependency on moneylenders for productive and consumptive loans decreased by 50%. • 1/3rd of the total household income generated through on-farm (poultry, dairy, goatry, etc.) and off-farm (brick/basket/leafplate making, etc.), activities as supplementary to agricultural income.	• Mid-term evaluation to see the trends. • Post-project evaluation	• Seasonality of food availability with each S-E classes • Income and expenditure analysis of each S-E classes • Matrix ranking of different crop and production, etc.

Social

• Level of awareness on social issues increased, and women and poor are given due share in the development process of the society.	• The rate of education among children, in particular, girl child increased from X to Y; • School drop out rate minimised from X to Y; • Awareness level increased for the importance of vaccination and family planning; • Awareness increased about health, education and other services, and quality services ensured; • Reduced level of liquor consumption; • Decreased number of community conflicts and growing mechanisms for resolving community conflicts internally without external interference; • The drudgery faced by women for fetching drinking water, fuel wood and fodder is reduced.	• Survey • School record • Record of Health Department • FGDs observation/govt. record • FGD market survey • FGD/police records • FGDs with women • FGDs/available records with WC/PIA report	• Survey or secondary sources • Fads • Chapati diagram • A day in women's life • Seasonal working calendar of women • Stakeholder analysis

(Table 5.3 Continued)

Outcome Results to be Achieved Out of the Activities	Success Indicators	Means of Verifications	Participatory Methods
Institutional •Sustained community action for the operation and maintenance of assets created and further development of the potential of the natural resources in the watershed	•All assets created under watershed are maintained and benefits accrued in the same proportion. •WDCs have mobilised additional resources from line departments for further development of natural resources. •WDC is financially sustainable to carry out maintenance work. •X% of credit needs of the community are met through SHG sources and Gramkosh. •Adequate availability of financial resources through local mechanisms (SHGs, Gramkosh) to meet at least X% of the average credit needs of the hh. •Marketing system in place for agricultural production to minimise the effect of middlemen. •Increasing trend in WC, SHG and UGs from observer-discussant-decision-maker. •Women and poor have access and control over the benefits accrued through watershed activities.	•Mid-term evaluation to see the trend •Post-project evaluation •Periodical reports by PIA •Concurrent monitoring reports by DRDA/PIA •Reports from Banks/LAMPs, etc.	•FGDs on past experience of maintenance of common property •Chapati diagram to understand institutional linkage of the village •Credit need analysis including sources and seasonality of need •FGDs on marketing mechanism and pros and cons

Source: Collector, Jhabua (1994)

Table 5.4 Activities Analysis

Activities		
Activities (What we do)	Success Indicators	Means of Verifications	Participatory Methods

Activities (What we do)	Success Indicators	Means of Verifications	Participatory Methods
Environmental			
•Biomass production increased from X to Y	•At least 50% of the total rainfall and 80–90% of the available run-off in the watershed is conserved.	•Action plan	•Resource mapping
	•Around 80% of the total land area of the watershed is treated under various SMC measures.	•Post treatment resource mapping	•Transect
	•Around 80% of the common property resources (revenue/forest/panchayat land) are taken over for operation and maintenance.	•PIA report—physical and financial	•Timeline on CPR
	•At least 1/3rd of the watershed area is covered under vegetation through plantation and natural regeneration.	•Observation during field visit	
		•Concurrent evaluation	
		•Remote sensing	

(Table 5.4 Continued)

Activities (What we do)	Success Indicators	Means of Verifications	Participatory Methods
Economic	•Double cropping area increased from X to Y with special reference to the land of BPL families •*Rabi* irrigation potential increased from X to Y and survival irrigation for *Kharif* increased from A to B •X% change in cropping pattern from food crop to commercial •Cropping intensity increased from X% to Y% •Appropriate land management practices adopted as per the gradient and other attributes of the land X ha. Under *silvi-pasture* - X ha. under *silvi* horticulture: X ha. under agro-forestry/ horticulture •At least X% of the households successfully adopted livestock breed improvement programme •X% of the households is engaged in off-farm activities like brickmaking, bamboo, NTFP, etc.	•PIA report •Post treatment resource mapping •Report of agriculture dept •Remote sensing •Physical survey	•Mapping of cropping area (*kharif, rabi*), irrigated, unirrigated with sources •Matrix, seasonality, pie diagram, survey, interview, FGDs to understand the base line conditions, land, water, vegetation and livestock

Social		
• Awareness camps organised on issues like education, health, hygiene, etc. on a regular interval.	• PIA report	• Social mapping
• The regular visits of health workers, teachers and other department's staff ensured.	• Field visits	• Role analysis of women
• At least 80% of the women are organised into SHGs (*Baira ni Kuldi*.)		• FGDs
• Women's active involvement in the planning, execution and monitoring of the work is ensured.		
• Opportunities created for the women to express their views.		
• Campaign against liquor organised at regular intervals.		
• Drudgery reducing technologies/ interventions introduced to facilitate women to take part in their strategic role.		

(Table 5.4 Continued)

Activities (What we do)	Success Indicators	Means of Verifications	Participatory Methods
Institutional	•All households in the watershed are the member of the Watershed Association.	•WC records-membership register, meeting proceedings, work register/ Gramkosh register, etc.	•Social mapping
	•At least 80–90% of the households (including poor and women) have actively participated in the planning process of the watershed programme.		•Wealth ranking
			•Chapati diagram
		•PIA report	•Livelihood analysis
	•At least 80–90% of the hhs (including poor and women) have been given exposure to successful watersheds.	•SHG records including bank pass book	•Time line
			•Trend matrix
	•At least 50% of the hhs (including not less than 75% of the BPL families) have received skill training–technical and managerial.	•Field visits	•Flow diagram
		•Review of action plan	•Pie diagram, etc.
	•At least 80–90% of hh are organised into SHGs and 80% of the hh into UGs. SHGs generated at least X amount of savings and have rotated as credit among themselves.		
	•At least 70–80% of the SHGs have mobilised X amount of external resources (NABARD, SJSY, etc.) as for meeting larger productive credit needs.		
	•At least 30–40% (average) of the cost of the physical work is contributed by the community and constituted Gramkosh and development fund.		

Institutional	
	• All Community organisations (Watershed Committees (WCs), SHG, UG) have developed and used their own functional norms.
	• WCs have actively participated in the implementation of the action plan.
	• WC have developed working modalities and got it approved by the Watershed Association (WA).
	• WC demonstrated its accountability to the WA and larger Gramsabha of the panchayat by creating transparent systems of work approval, financial transactions, etc. (viz. the treatment plan approved by the WA may be painted on the wall of a common place like, panchayat/ school building; All financial transactions, of a month may be read out in the monthly meeting of the WA, etc.
	• WC developed norms for the use of Gramkosh and also the distribution of benefits.
	• UGs/WC have maintained all the structures created under watershed work.
	• Services of line departments are converged at the watershed level.
	• Sustained collective community action for watershed plus interventions like co-operative mkt, input supply, etc.

Source: Collector, Jhabua (1994)

This project also envisaged the impact change that would be brought about as a result of the implementation of the project (see Fig. 5.3).

Fig. 5.3 Impact Change Diagram

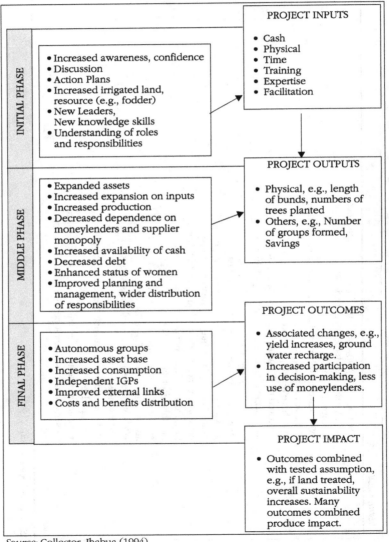

Source: Collector, Jhabua (1994)

Monitoring and evaluation

This project also provided for monitoring and impact assessment in order to periodically evaluate the progress in its implementation and the results achieved. Provision was made in this project for collecting details on parameters such as groundwater table and well density through observation wells, which were to be identified following standard procedures. Cropping patterns and crop yield were to be addressed periodically, based on observed changes in these indicators. It also provided for intermediate assessment of the progress achieved in its implementation, the objective being to determine the impact of this project in accelerating economic growth, providing basic needs and maintaining ecological balance. The intermediate assessment was expected to assess the role of NGOs and government functionaries in creating awareness among people. It was also required to analyse constraints, if any, in implementation and issues related to the social acceptability of watershed development.

Using techniques of remote sensing in randomly selected areas this project also envisaged impact evaluation. The satellite imageries were expected to show pre-watershed and post-watershed activity status, particularly in respect of increase in *rabi* areas, formation/augmentation of waterbodies and increase/decrease in wastelands and fallow lands.

Analysing this project

Was this project sound? The following analysis looks at its structural, managerial, planning and technical aspects in addition to discussing how far it succeeded as an empowering strategy in its formulation.

Structurally, it managed to capture almost all the elements that a successful project ought to have. It propounded a vision that was forward looking, and its charter was comprehensive, encompassing as it did, environmental, economic, institutional and social perspectives. The intended outcomes were related to the elements in the vision statement. This model had unambiguous forward and backward linkages, and its implementation led to the realisation of its vision. Where the project was deficient structurally was in dealing with the exit protocol. It was very likely that conflicts would arise between the Watershed Development Committees and the Gram Panchayats after the project period was over. It did not provide any mechanism for conflict resolution. Perhaps this project could have provided for an institutional mechanism by way of conflict resolution: the Watershed Development Committee and the Self-Help Groups could have been made members of the Gram Panchayat. Another means of resolving conflicts could have been by way of proposing amendments in the revenue and panchayat laws to make the concurrence of the Watershed Development Committee mandatory in respect of certain issues like mining, liquor auctions, etc.

Managerially, this project proposed suitable administrative and institutional arrangements. In their design, these arrangements provided for a multi-tier structure with specified roles and responsibilities for each player. The primary players involved were the district administration, the PIAs, the Watershed Development Committees, Self-Help Groups, user groups and the people as participants. The structure of the implementation and monitoring as proposed by it underscored the criticality of various levels of the institutional arrangement working in tandem. In fact, what was required for the successful implementation and monitoring of a project was multi-layered, as the watershed project involved both vertical coordination of different tiers of district administration and horizontal coordination across the various agencies at the level of micro watershed. The design of the watershed project took care of that.

From the perspective of planning, it broke new ground by using logical framework analysis as a tool for planning at the micro level. What it did was that it set out success indicators and means of verification in clear and identifiable terms, participatory methods being the common referent. Apart from validating the assumption that logical framework analysis can be potentially used for planning and monitoring micro-watershed programmes, it established a functional relationship with the elements of the project vision by setting out the physical, social and economic activities, targets and outcomes with indicators. The use of a sophisticated planning technique for micro-level project in a rural setting is indeed laudable.

From a technical perspective, the objective of a watershed project is to generate optimally suitable action plans for a geographical area so as to harness the productive potential of the local resources to sustainable levels. While doing so, it should incorporate contemporary technologies compatible with natural resources endowments, socio-economic profile and climatic parameters. In the formulation of the watershed project in Jhabua, the integration of these aspects was proposed to be achieved systematically through a combination of scientific tools, people's aspirations and local wisdom. The project, in fact, envisaged the preparation of resource maps, water maps, land use maps and geographical maps. Such mapping was expected to encapsulate sustainable agricultural practices, soil and water conservation measures, land conditions and needs. It expected local knowledge and wisdom to contribute vital inputs to the preparation of these maps.

In its conceptualisation, it was basically sound in terms of problem diagnosis and identification. Social issues were identified, indicators were provided and an appropriate strategy was delineated. In terms of economic issues, this project provided both indicators and strategy. In respect of biophysical issues, problems were diagnosed and identified, and the treatment strategy was indicated. There was a fair amount of risk analysis and drought-proofing measures were offered as a solution. The project also suggested convergence of activities at the grassroots level, particularly in terms of actions by the line departments of the government. Resource requirements were indicated, and cash flows provided for. There were suggestions for periodic appraisals, monitoring, analysis and evaluation.

Institutional context

We had pointed out in Chapter 4 that an appropriate institutional context should be established for the empowerment strategy to succeed. We had talked about suitable reordering in the structure of both formal and informal institutions that affected the lives of the poor. In the institutional context, we had talked about social and cultural aspects that are particularly important for empowerment exercises: while building assets and capabilities, attention is to be paid to the fact that institutional structures reflect local norms, values and behaviour, and that awareness-raising should be done through participatory methods, in which local knowledge based on experience should be combined with outside knowledge. We had also talked about the criticality of the role of civil society intermediaries in empowering the poor during the project implementation period and enabling them to sustain the process after the project was completed.

This project did, indeed, set an appropriate institutional context for empowerment of the Bhil. In the first place, by suggesting the creation of community organisations that are participatory, institutionally sustainable, capable of identifying and addressing common development issues, truly democratic in their functioning and sensitive to the causes of the poor and women, it created conditions for establishing the institutional context for empowerment. It also suggested an increasing trend in the functioning of these community organisations: the participating Bhil were to be enabled, over a period of time, to graduate from being observers to being discussant and ultimately, to becoming decision-makers.

The social and cultural aspects of the institutional context were also addressed by this project. For example, in terms of its stipulations, it was the village community that was required to prioritise problems and identify solutions. Significantly, it was a critical component of the watershed project that the Bhil themselves prepare the action plans, their costing and how the activities should be scheduled. Since the entire village community was to be involved in the process of problems prioritisation, solution and identification and preparation of the action plans and its costing, the traditional knowledge and wisdom of the Bhil community based on its experience were expected to be gainfully used in the planning of the watershed activities.

The association of the civil society intermediaries (NGOs) in the conceptualisation, formulation and implementation of the watershed project and action plans was a key component of this project. The NGOs, which were the Programme Implementing Agencies (PIAs) in charge of each milli watershed, were made responsible for sensitising the participating Bhil and mobilising them into community organisations. This is a difficult task in the Jhabua setting where the Bhil shun contact with the outside world and are sceptical of any external initiative. So their confidence had to be won. This project contemplated the NGOs to take up entry point works such as construction of bus stands, school buildings, repairs to temples, sanitation works, health camps, repairs to hand pumps and horticultural crops. It also envisaged that through these entry point works, the NGOs would build rapport, win the trust of the Bhil and get accepted by them.

It envisaged the association of the NGOs in many other ways. They were required to cultivate in the Bhil a sense of self-reliance and self-confidence and organise them into viable groups. They had to provide technical assistance and skill building exercises to make these groups sustainable and effective. They were called upon to help the groups in scaling up their membership, range of functions, awareness raising and training. They were also required to create bridges between these Bhil groups and the departments of the government and outside agencies.

On the whole, the watershed project planned for putting in place an appropriate institutional context which would give the Bhil control over institutions that affect their lives while also giving due consideration to its social and cultural aspects. Civil society intermediaries were given strategic roles in empowering the poor during this project implementation period and create conditions for sustaining the process after it was over.

Empowerment framework

Does the project provide an appropriate framework for empowerment? In Chapter 4, we had suggested an empowerment framework which diagrammed the relationship between institutions, empowerment and improved development outcomes for the poor. Also listed out were the development outcomes that would flow as a result of reforms in the institutions and investment in poor people and their organisations, which included: improved governance, inclusive basic services, strengthened civil society, strengthened poor people's organisations, and increased assets and freedom of choice.

The way this watershed project was designed, improved governance was certainly an intended outcome. It emphasised provision of inclusive basic services; in fact, it contemplated that as a result of the awareness camps organised under the aegis of the project, the Bhil themselves would demand and ensure regular visits of health workers, teachers and other government staff, and ensure the quality of basic public services.

On the whole, the way it was designed had the potential for enhancing the capacity of the Bhil to influence the institutions of state and society by strengthening their participation in processes of local governance and decision-making. It also delineated a suitable framework that was capable of generating developmental outcomes to empower the Bhil.

6

The Action Plans

Participation of the Bhil was considered the key component of this project. This was a difficult task in the Jhabua setting where the Bhil shun contact with the outside world and are sceptical of any government initiative. So their confidence had to be won. For this purpose, the PIAs took up entry point works such as construction of bus stands and school buildings, repairs to temples, sanitation works, health camps, repairs to handpumps and horticultural crops. Through it, the PIAs hoped to win the trust of the Bhil and get accepted by them.

The next step was to educate them about the importance of watershed and create awareness. This was done by organising rallies, meetings, Kala Jathas and street plays on the theme of watershed development. The contents of these were developed by the Madhya Pradesh unit of the Bharatiya Gyan Vigyan Samiti, which depicted elementary concepts of soil and water conservation and watershed management in a language and manner that the illiterate Bhil could understand. The take-off level had to be kept low, keeping in mind the capacity of the Bhil to comprehend. Exposure visits were organised for the members of the Self-Help Groups, user groups, Watershed Development Committees and field-level staff of the PIA to places outside the district such as Ralegaon Sidhi, Urli Kanchan, and Alwar, where successful watershed projects had been implemented with people's participation.

This project started with Participatory Rural Appraisal (PRA), a process in which the PIAs, the Watershed Development Committees, Self-Help Groups and user groups get an appreciation of the watershed resources, problems and possible solutions. This is followed by the preparation of an action plan, its costing, and scheduling of the activities.

The following steps were followed in the PRA:

- Social Mapping—positions of houses, roads, temples, handpumps, caste groupings.

- Land Use Mapping— survey numbers, cropping pattern, productivity status, irrigation facilities, fertiliser use, pesticide use, organic manuring.

After the preparation of the social and land use maps, following maps were prepared.

- Geographical Map—topography, soil texture, environmental problems (water-logging, soil erosion, seasonal flooding, salinity), agriculture, horticulture, forest area and wasteland.
- Water Map—tank, tubewell, river, handpump, siltation status of river and pond, dam, bund, drainage line, water-logged area, water quality, ownership, users of water resources, drainage point, water table and irrigation methods.
- Seasonal Map—rainfall month, migration cycle, fodder availability, timings of diseases, availability of drinking water and cropping cycle.
- Time Line—records of basic historical facts, for e.g., certain important technological developments, productivity of land, rulers, population changes, infrastructural facilities and forest status.

It was assumed that while preparing the resource maps, water maps, land use maps and geographical maps, the villagers would get an understanding of the problems of the watershed. Then, these were prioritised on the basis of the needs of the village and the intensity of the physical–environmental damages. After that, the action plan was prepared, its costing was done, and the activities proposed in the action plan were scheduled in a time sequence.

All the steps of the PRA process were followed while preparing the action plans for the watershed project in Jhabua. As a first step, the staff of the PIAs was trained in PRA techniques, and workshops were conducted at the district level. They were educated about the natural resources of the village, the social structures, and the need for people's participation in this project; they were also given the necessary skill in natural resource mapping, social mapping, time index techniques, preference racking and quadrant matrix. Extensive land-literacy campaigns were then launched to get an idea of the watershed resources, problems and possible solutions. The members of the Watershed Development Committees, the staff of the PIA, and the members of the Self-Help Groups and the user groups were involved in the land-literacy campaigns.

The first step was social mapping. Maps indicating the positions of the houses, borewells, open wells, streams, roads and agricultural fields were plotted on a map in respect of each *falia*. Below is a social map prepared at the time in respect of Gopalpura (Basunia *falia*) (see Map 6.1).

Other maps such as the resource map, water map, land-use map and geographical map were prepared using the PRA techniques. Time-line was also prepared, depicting the basic historical facts and

Map 6.1 Gopalpura (Basunia *Falia*)

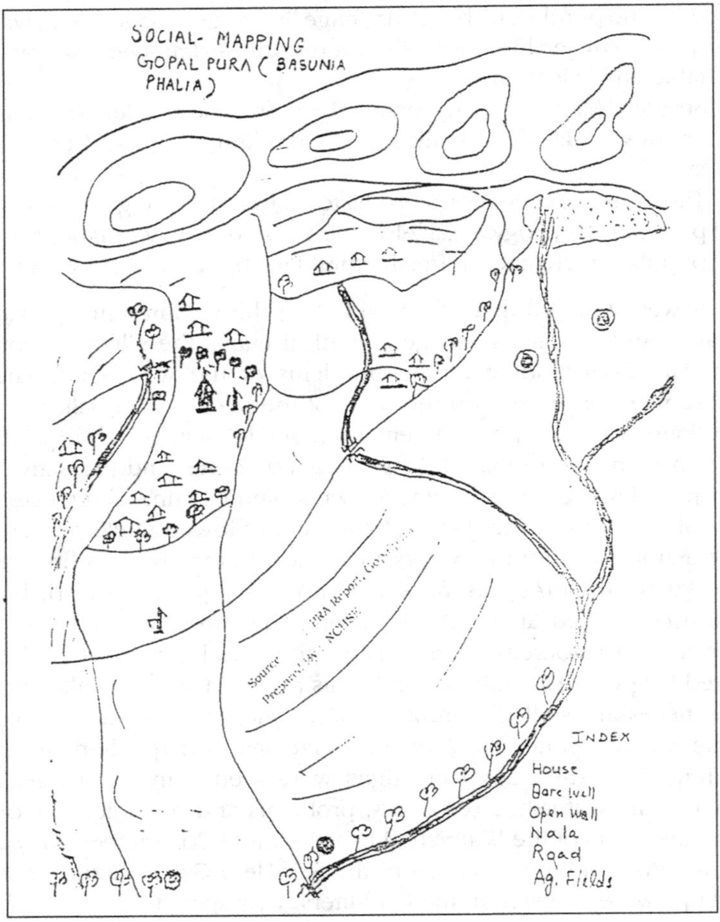

Source: Mahajan (1996)

the changes over time. Here is a typical example of time-line prepared for Gopalpura village.

Time Line for Gopalpura Village

1920–30 : Raja Uday Singh was the ruler.
He and his family used to come to Bhagon and stay for months.
Raja Uday Singh went hunting; and the forest then was dense.
Villagers used to patrol the village boundaries by rotation.
Kind was used to hire labour without paying them
their due wages.
There were 12 houses in the village and silver coins were in use.

1940–50 : Rs 10 was the standard bride price.
Then turban was not in use and people used to wear *langoti*
(small indigenous under garment).
There were atleast 60–100 livestock in each household.
The church was constructed during this time period.
Gram, castor, maize and cotton were the main crops.
Availability of water was scarce.
*Once Mrs Tarej, a resident of Gopalpura village, fell ill.
Nothing seemed to work for her. Finally she prayed in
the church and was cured. This made her convert to Christianity.

1955–65 : The bicycle came to the village.
Roads were constructed.
By this period, 50 houses were constructed in the village.

1975–85 : People had started to fell trees.
A tank was constructed by the then government.
A temple was also constructed.

1990 : Villagers started to grow wheat.
Bride price now rose to as high as Rs 25,000.

1995 : There are 210 houses in all.
Electricity is supplied to individual households.
The once dense forest cover is stripped bare.

* Information collected from
Mrs. Tarej, Age – 80 years
Village : Gopalpura

Source: Mahajan (1996)

Such an exercise enabled the members of the Watershed Development Committees, the staff of the PIA, the Self-Help Groups and the user groups to ascertain the geographical, geo-morphological and geo-hydrological problems of the village. House-to-house contact and PRA techniques were helpful in identifying the social and ecological problems and listing the perception of the villagers. Inventories were made of the watershed resources, problems and possible solutions.

The villagers were now in a position to understand and appreciate the problems of the watershed. After this, the village community proceeded to prioritise problems, identify solutions, and prepare action plans. It was a critical component of this project that the people themselves prepared the action plan, its costing and the schedule of activities. Since the entire village community was involved in the process of problem prioritisation, solution identification, preparation of the action plan and costing, the traditional knowledge and wisdom of the Bhil community could be utilised in the planning of watershed activities.

A large number of training programmes and workshops were conducted as part of this project. They covered themes such as utility of remote sensing, watershed, PRA and RRA techniques, preparation of action plans for watershed development, resource mapping, project management, maintenance of accounts, tree plantation and grassland development, agricultural techniques in watersheds, fisheries and poultry development, credit mobilisation for Self-Help Groups, book keeping and ledger management, etc.

The participants in these training programmes and workshops were the staff of the PIAs and multi-dimensional task groups, members of Self-Help Groups and user groups, chairmen, secretaries and volunteers of watershed development committees, the project officers and members of the project team, and officials administering the project. Exposure visits were organised to places within the district and outside, such as Ralegaon Sidhi, Urli Kanchan, Alwar, etc. Villagers and staff of PIAs were taken to Godhra in Gujarat on tours of the successful horticulture programme. They were also taken on educational tours of Rajasthan to familiarise them with the watershed experiments in that state.

Some of the workshops and training programmes were conducted outside the state with the help of organisations which have successfully implemented participatory watershed programmes. For example, Tarun Bharat Sangh of Alwar, Rajasthan conducted a workshop on the construction and maintenance of community-owned tanks. Training programmes for the volunteers of the Watershed Development Committees were conducted at Dahod in Gujarat.

The training programmes, workshops and exposure visits achieved two important objectives. First, the process educated the Bhil about how watershed management was important in a rural setting with agriculture as the only source of livelihood. It informed them, in a manner they could understand, about the role played by topsoil and soil moisture in plant growth, criticality of forest and vegetation in soil and water conservation, long-term ill effects of un-controlled grazing and the need to conserve pastures, and the system of movement of water through the watershed.

Second, it made the Bhil see that the participatory watershed pro-gramme has in-built structures to assure the participation of bene-ficiaries at every single stage of decision-making. It told them how they were involved in planning the watershed activities, how they had direct control over financial management, and how their own contributions were utilised while implementing this project.

As we noted, a land literacy campaign was taken up as a first step and an estimation was made of the required structures of soil and water conservation. In consultation with the Self-Help Groups, user groups and Watershed Development Committees, an action plan was prepared for each micro watershed.

Soil and water conservation

While preparing the action plan, it was recognised that for con-serving soil and water, the quantity and speed of water which flows on the surface had to be controlled, and all measures that reduced the quantum of surface run-off and its speed were to be adopted in conserving soil and water. For this purpose, this project proposed ridge to valley treatment. Treatment types were also suggested, depending on the topographical zone of the watershed. While pre-paring the action plans, three zones were recognised in the watershed area: Upper area or recharge zone; Middle area or transition zone; and Lower edges or discharge zone.

In the action plan, the following treatment for the recharge zone was suggested: contour trenching; contour bunding with vegeta-tive protection; grass cultivation and vegetation with controlled grazing; afforestation (e.g., bamboo, Seevan, Subabul, Julliflora, Acacia etc.); bench terracing on steep slopes; water outlet channels on the hills; and contour farming. While in the middle or transition zone, the following handling was suggested: farm bunds; check gully erosion; underground *bandharas* to check sub-surface water flow; wasteland development; and pasture development.

For the lower edges or the discharge zone, the action plans sug-gested the following measures to be taken up: farm ponds; percolation; check dams; water harvesting structures/ dams; and field mulching and organic manuring.

Regeneration of degraded forests

As rapid loss of forest cover was seen as the most serious environmental threat, the priority area for consideration was regeneration of degraded forests. In this scheme of Joint Forest Management (JFM) (already introduced in the district), the government and the local community join hands to protect and manage the forest resources. In this partnership, each partner is given a specific role and the benefits are shared. While planning for the regeneration of forests, the project followed the pattern laid down in the JFM scheme. In the action plans for watershed activities, the JFM scheme and the watershed works were made complementary to each other ensuring that both work in tandem to regenerate the forest. They provided for natural restoration from dormant seeds, coppices and gap planting in the watershed areas in collaboration with the village forest committees established under the JFM scheme.

Plantation activities

The action plans proposed planting of trees in both private and community lands. For this purpose, soil-working and seed-sowing were planned over large areas. Plantation of ratanjot, sitaphal and moringa trees were proposed through seed sowing on a large scale, while those of fruit-bearing species like mango, amla, ber and guava were planned over privately owned lands. Multipurpose trees like *samel, jungle jalebi*, bamboo, *safed khair, babool*, eucalyptus, citrus lemon were also proposed. While planning plantation activities, care was taken to see that only those species were selected which could thrive in the dry climate of Jhabua district. Chosen species included fuel wood trees as well as trees from which the Bhil could get building materials and forest produce. High-density energy plantations of safed khair and babool were planned over community lands.

As the departmental nurseries of the Forest Department or the Horticulture Department were not in a position to cater to the demand of the implementation of such large-scale plantation activities, it was decided to decentralise nursery activities, hence plans to establish nurseries at several levels. Under the kitchen garden scheme, villagers were encouraged to establish it in lands available in and around Bhil houses. For households below the poverty line, concessions were offered.

Members of Self-Help Groups who were interested in raising nurseries were identified and trained. They raised nurseries of 15,000 to 1,00,000 by availing of loans from funds available with the watershed development committee. At the level of the milli watershed, larger nurseries were planned—for species such as bamboo, *amla* and fruit-bearing trees. The PIAs were given the responsibility for such nurseries with 2 to 5 lakh plants; money for which was proposed to be given from the funds available with the PIA.

Fodder development

Development of fodder was an important item in the action plans; it was planned both in privately owned and government lands. It was assumed that fodder development schemes would provide the environment for biomass generation and soil conservation; in fact, while planning soil and water conservation measures, it had been suggested that pasture development and *silvi pastoral* activities should be taken up in the transition and discharge zones of the watershed. It was assumed that the watershed villages would become self-sufficient in fodder and forage by the time of the completion of the project, and that fodder development would provide immediate returns to the villagers.

The action plans provided for the village community to take up grassland development in common land. It also detailed how the grass harvested should be shared by the village community. The benefit-sharing mechanism was set out as in the following.

- Time Sharing: a day to be specified for a particular member of the community. On that day, the family cuts grass according to its capacity and keeps it.
- Place Sharing: the place for each family is specified. The family has the authority over the grass of that place.
- Produce Sharing: all the people in the community cut grass, which is then collected in the village and distributed among villagers by the Watershed Development Committee on the basis of the estimation made by the committee of the requirement of each family.

There was even a stipulation that 50 per cent of the grass which is gathered from the common land should be sold at the market price and the proceeds deposited in the Gramkosh.

Certain varieties of fodder crops such as *Hamata, Dinanath, Sukli* and *Batodi* were proposed in the action plans, although the actual decision on what fodder crop to grow was left to the villagers. They were also encouraged to grow fodder crops along with their regular agricultural crops. For policing the fodder crop grown in the common land, the system provided in the JFM scheme was adopted. The idea was one of social fencing—the voluntary protection of the crop by the villagers with minimum fencing—which contemplated that the villagers would form protection committees with membership from all the families. This meant that the fodder crop would be protected by the members of the committee on a rotation basis. The idea of an effective social sanction mechanism in the form of fines and punishment for illegal intrusion was mooted. The fines collected were to be deposited in the Gramkosh.

On the whole, the idea of participatory management of fodder cultivation by the village community was proposed. It was assumed that when social fencing becomes an effective way of protecting the fodder crop, there would be no need to erect other types of fencing like stone wall, cattle-proof trenches and biological fencing, and as a result, the cost of such physical fencing could be a net saving to the village community.

Wasteland reduction

The reduction of wasteland was an important part of the action plans. It was supposed that soil and water conservation measures such as gully plugs and staggered contour trenches would increase soil accumulation in the wasteland, while a combination of social and vegetative fencing and physical measures such as cattle protection trenches and stone bunds would provide the necessary protection.

Water harvesting tanks

The action plans projected water harvesting tanks to be constructed in the discharge zone of the watershed. They were to be built by the village community. The idea was that the villagers would select the sites, build the tanks using local knowledge and wisdom, and manage both the storage of water and its utilisation. The cost of constructing the tanks was to be borne partly by the villagers and partly from the funds kept at the disposal of the watershed development committees. It was assumed that at least 50 per cent of the cost would be met by the villagers by way of voluntary service or contribution in the form of labour and material. The responsibility for the construction and maintenance was given to the watershed development committees, while the management of the utilisation of water was with the user group.

Groundwater resources

It also proposed that while the use of surface water could be the primary basis in the watershed development strategy, the groundwater should be conserved for selected purposes only. Upgradation of groundwater resources was recognised as an important objective while preparing the action plans. Percolation tanks were proposed to be taken up in different micro watersheds. They were expected to recharge groundwater and also make water available for watering of the livestock. It was assumed that due to the increase in sub-surface moisture, there would be vegetation surrounding the percolation tanks and an increase in lift irrigation schemes as a result of enhanced water availability. Sub-surface dykes were also planned; the idea was to recharge groundwater. These structures were proposed to be located in the transition and discharge zones of the watershed, and on sites identified through remote sensing maps and village transects.

Increase in water resources

It was assumed that, as a result of the measures planned in the watershed, there would be a significant increase in water resources. This would be due to enhanced recharge in the upper catchment and prolonged flow of surface water as a result of soil and water conservation measures and construction of stop dams. It was necessary, however, to measure the increase in water resources. For this purpose, a number of observation wells were planned in each micro watershed to record the water level in the observation wells on a regular basis. It was also proposed that the people's perception of the period of flow in streams should be gathered in group discussions and recorded, so that it could be used as a benchmark for plotting the increase in water resources.

Nadep and biogas

It was assumed that with the enhanced availability of fodder, the number of large ruminants in the watershed area would increase. Both the harvested fodder and crop by-products would be used for feeding ruminants, and this would mean an upgradation in the potential for recycling nutrients. The action plans, therefore, proposed that the dung produced should be used to make compost; in other words, the creation of a good organic fertiliser resource in the form of Nadep. They are aerated compost tanks; it is a method by which a Gobar Khad is prepared in a simple technique—as simple as handling a routine household work. In the action plans, villagers owning a number of cattle heads were selected for the Nadep and biogas programmes.

Grain banks

The action plans proposed that grain banks should be established at the village level; the idea was to improve the level of nutrition and provide foodgrains through networking, and more importantly, to relieve the Bhil from their dependency on the village moneylender in the lean months before the *kharif* crop was harvested. It also meant that, with the establishment of grain banks, the village community would have taken the responsibility of looking after its starving and indigent members at times of scarcity, while, at the same time, reinforcing the feeling of self-help. The money for setting up the grain banks was taken from the funds meant for entry point works. The responsibility for the establishment of the grain bank was given to the Watershed Development Committee. They had to select a safe building for the location of the grain bank and inform all the families in the village about the grain bank so that they could make use of facilities provided by the bank.

All the families in the village were to be made members of the grain bank. The bank could advance 50 kilos of food grain to each family in two instalments. The Watershed Development Committee was also given the power to make exemptions in suitable cases and provide a maximum loan up to 100 kilos to a needy family. An active working group which would maintain and manage the grain bank was also proposed. The volunteer of the watershed development committee was appointed as the ex-officio member-secretary of the working group. Seven members were to be elected to this group—half of them women—and the life of the group was for a period of three years. The group was supposed to keep a tab on the quantity of foodgrains in the bank and the rate of interest to be charged, prescribe a time-frame for the return of the loaned foodgrain, and maintain the records in respect of each loanee-member. The active working group was to meet at least twice a month to transact its business.

Guidelines were prescribed for the rate of interest that could be charged:

- If the foodgrains are returned within three months, the rate of interest should be 5 per cent.
- The rate of interest should be 10 per cent if the foodgrains are returned between three to six months.
- If the foodgrains are returned after six months, the rate of interest should be 15 per cent and the penalty should be as decided by the active working group.

Falia schools and poriawadis

Jhabua district, as we noted earlier, has a peculiar human settlement pattern, with each village consisting of several *falias* spread over a radius of 4 to 5 kilometres. Each *falia* does not have a school, and the children have to cover long distances to go to a school. In order to cover the children in the dispersed *falias*, the action plans mooted the idea of Self-Help Groups opening schools and institutions of pre-primary education; the *falia* Schools and poriawadis. The Self-Help Group was expected to pay for the salary of the teachers and maintain the schools.

An empowering approach

The question is: How far did the process of preparing these action plans incorporate an empowering approach? As we have seen, while preparing the action plans, care was taken to see that the Bhil were educated about the importance of watershed in order to create the necessary awareness. This was done through rallies, meetings, Kala Jathas and street plays, which depicted elementary concepts of soil and water conservation in a language and manner that the Bhil could understand. In addition, while preparing the action plans, extensive land-literacy campaigns were taken up to give the Bhil an idea of the watershed resources, problems and possible solutions. Exposure visits were also undertaken for the members of the watershed development committees, Self-Help Groups and user groups to places outside the district where successful watershed projects had been implemented with peoples's participation. In other words, the preparation of action plans was based on seamless information flows, and it was expected that armed with such information, the participating Bhil would be in a position to understand the problems of the watershed, prioritise problems, identify solutions and prepare action plans.

Regarding inclusion and participation, all the participating Bhil households in the watershed area were included in the task of preparing the action plans and actively participated in their preparation. The fact that Participatory Rural Appraisal (PRA) methods were used in preparing the action plans is a testimony to this. Though the PRA process, the participating Bhil households got an appreciation of the watershed resources, problems and possible solutions. In addition, the Bhil households were also associated with the preparation of the resource maps, water maps, land use maps and geographical maps which were done by following the PRA process. Since the entire village community was involved in the process of mapping the watershed, prioritising problems and identifying solutions, the traditional knowledge of the Bhil community could be utilised in the preparation of action plans.

In respect of accountability, the association of the Bhil with the preparation of the action plans made them realise how a participatory watershed project incorporates in-built mechanisms to ensure the participation of the Bhil at every single stage of decision-making. It made it clear to them how they were to be involved in the implementation and monitoring of the action plans and how they were given direct control over the financial management of the project. It also told the participating Bhil how they were given the responsibility of managing community institutions such as the Self-Help Groups, user groups, grain banks and watershed development committees, and how through the management of such institutions, they were put in a position in which they were suitably empowered to call government functionaries and providers of public services to account by holding them responsible for their policies, actions and use of funds.

The local organisational capacity of the Bhil was tried and tested in the very act of collectively preparing the action plans. Local organisational capacity is about the ability of the people to organise themselves, work together and raise resources to solve problems of collective interest. The fact that, in the process of debating and finalising the action plans, the Bhil could prepare the necessary maps, prioritise problems, identify solutions and prepare action plans with its costing and scheduling of activities, gave the necessary encouragement to the abilities of the Bhil to organise themselves, work together and deliberate to solve problems of collective interest. This was a definitive exercise in bonding social capital of the Bhil at the village level.

On the whole, the preparation of the action plans for the watershed project in Jhabua incorporated all the elements of an empowering approach.

7

Civil Society Intermediaries

The watershed project in Jhabua also delineated a very important role to the civil society intermediaries. Their initial task was to bond social capital within the village community, and towards that end, they had to win the trust of the Bhil and gain acceptance. These civil society intermediaries also had to organise the Bhil into viable groups, and take on the critical position of supporting the Bhil, translating and elucidating information about government schemes and programmes to them, and thus help in linking them to governmental institutions. In other words, the NGOs, as the PIAs of the watershed projects, were to be instrumental in empowering the Bhil.

Here is an attempt to draw a profile of two NGOs who worked as PIAs for two most difficult areas of the watershed project. What follows is an analysis of their role and responsibilities in those areas and of the extent they discharged them. This analytical study is entirely based on interactions with them.

Action for social advancement (ASA)

ASA, an NGO, is the PIA for 25 villages covering an area of 10,000 hectares in Udaigarh and Jobat blocks. This NGO has a team of 54 development workers—of these, 19 are professionally qualified in fields like rural development, social work, forestry management and engineering. There are 31 field workers; in addition, there is a support staff of four.

Mission

ASA was founded by a group of professionals who had been working in the Bhil area since 1996. Their first-hand experience gave them an understanding of the underlying causes of the problems of the Bhil area and the patterns in which they were manifested. Their mission is to conduct an extensive and intensive participatory development process through the empowerment of the Bhil community, with a special focus on the role of women and the socio-economically deprived sections. Replenishment of natural resources and initiating processes for sustainable management of natural resources has also been ASA's major planks.

Strategy

ASA's strategy has five components:

1. To create awareness among the Bhil as to enable them to understand the root cause of their various problems and to instil in them the confidence in their ability to sort them out.
2. To organise the Bhil into groups that can act as agents for change.
3. To increase the capacity of the Bhil to improve the quality of their life through various interventions.
4. To act as a demand creator for the groups so that maximum benefits/ resources, both internal and external, are generated.
5. To act as a facilitator to address the development needs identified by the Bhil themselves.

Community organisation

The organisation of the Bhil community into an aware and self-reliant institution that can take care of the community's interests is one of the major goals of ASA. It has effectively organised the Bhil community into various community based organisations which include Self-Help Groups, user groups and watershed development committees. According to ASA, Self-Help Groups are the functional units of the Community Based Organisations (CBOs), as they form the nucleus of all developmental activities including planning, implementation and review of the watershed programmes. Their policy has therefore been to arrange the existing kinship groups into SHGs.

According to this NGO, moneylending is one of the most thriving businesses and debt traps in the Bhil area. To counteract moneylending by professional moneylenders, the Self-Help Groups organised by ASA have adopted savings and credit as their core activity. While, on the one hand, such activities by the Self-Help Groups have ensured the smooth functioning and growth of the groups, on the other, they have generated resources for contingency requirements and investments, thus making the groups cohesive.

It has been ASA's policy that the Self-Help Groups should follow a single household representation norm wherein a household is represented by only one member. They have adopted this policy in order to ensure that the integrity of the households is maintained and that there is no likelihood of diffusion of responsibilities and conflict of positions.

ASA has also formed user groups, wherever necessary, for management of specific activities, but care has been taken to see that they are formed within the broad framework of the Self-Help Groups. These user groups are organised more as functional groups and they develop their own working norms. Most of them have been organised around water bodies created or rehabilitated by the watershed project.

According to ASA, village level institutions should be responsible for the management of wider development issues of the Bhil village such as watershed development or management of the forest resource. That is because they are usually the representative forum of the Self-Help Groups and the user groups. In the case of watershed development programme, it has been ASA's strategy that the Self-Help Groups should be made directly responsible for planning and implementation of the watershed programme while watershed development committees should act as the funding and monitoring agency at the village level. ASA has ensured that representatives of all the Self-Help Groups in the watershed area are there in the watershed development committee.

Micro credit

This NGO believes that savings and credit should be the core activity of the Self-Help Groups. ASA has ensured that every member of a Self-Help Group contributes a fixed sum at regular intervals. The general body of this group decides what the amount should be and what the interval should be. ASA has helped the Self-Help Groups in getting revolving fund from the banks (refinanced by NABARD), the watershed programme, the village development fund (Gramkosh) and the SHG federation for their credit activities. These groups provide credit to their members on consideration of creditworthiness. Productive loans are prioritised over consumption loans for social functions. It has been ASA's policy that the Self-Help Groups themselves should decide on the interest rate to be charged.

SHG federation

ASA has motivated the Self-Help Groups to come together in order to form a federation which will provide the group's support towards linkages with the government programmes, bulk supplies and purchase, provision of services and policy advocacy. The general body of the federation that ASA has helped form consists of two elected members of each Self-Help Group. The executive body is elected from each member Self-Help Group, while the president and vice-president are elected from the executive body members.

SHG federation

In its maiden effort, the federation at Bori arranged the supply of 1,700 bags of fertilisers to 32 of its member Self-Help Groups. The bulk purchase gave fertiliser to the Bhil at a considerably lower price. While the federation provided loans to eight Self-Help Groups, it also helped in linking seven Self-Help Groups in the NABARD-SHG scheme for availing of loans. The Gandhwani federation has linked with the Lead Bank of Jhabua district to avail of the loan facilities under NABARD-SHG scheme for 20 Self-Help Groups.

Gramkosh

In ASA's strategy, Gramkosh is one of the important sources for refinancing the Self-Help Groups. It is a village level fund formed out of the contributions of the Bhil from the payments they get for watershed development activities or any other user group activities. The Bhil contribute 50 per cent of their wages from private resource development activities and 25 per cent from the common resource development activities. ASA is of the view that such contributions make the watershed programmes sustainable not only by their ideology of participation, identification and ownership, but also by positively influencing the savings–investment ratio.

Contributions in the true sense

Theirs is the land. Theirs would be the benefits. So they 'should' and in the long-term only they 'can' make the investments. We at ASA don't have any illusions on this, nor do we expect that our people should have any. Contributions, therefore, are regarded as sacred.

(Box Continued)

> The people contribute 50 per cent of the labour cost for work on private land and 25 per cent on common lands. The amount is deposited in the Gramkosh from which people can meet their contingency and investment requirements. Initially, the WDC used to deduct the expected contributions at source. But soon this led to a number of misconceptions and didn't achieve the purpose. Of this realisation, a different procedure was born—now, full labour charges are paid to the people. They submit their contributions and get a receipt for that. This new approach has brought in the right attitude among the people, i.e., one of 'ownership'.

Natural resources development

According to ASA, the concept of land treatment in the form of watershed units has been highly successful in arresting the loss of soil and moisture in Jhabua. These measures have ensured conservation of soil and collection of rainwater, and such agronomic improvements have ensured sustained prosperity of the area. ASA has been intensely involved in the watershed development project, and it has acted as the facilitator and provider of technical inputs.

ASA takes a sociological approach to watershed development. The watershed guidelines issued by Goverment of India recommend micro watersheds managed by Watershed Development Committees as functional units of watershed development. Such an approach takes only the geographical aspects into consideration while it overlooks the social aspect, thereby making the process complex and difficult to sustain. ASA has followed the middle way by making the watershed development committee the larger forum representing all the Self-Help Groups of the watershed area while for all practical purposes a hamlet forms the watershed development unit. In their approach, there is no compromise on technical grounds as the physical measures remain the same, while it ensures sustainability as it complements and uses the existing institutions and practices.

Soil and moisture conservation

Private land

According to ASA, field bundings (earthen and stone) are a very effective method of soil conservation on cultivated, undulating land.

They are a part of the traditional land cultivation patterns, and ASA has made concerted efforts to ensure that this practice is used to its true potential. Binding grasses like style hamata have been planted over the bunds. The immediate impact of this activity is seen when the farmer starts taking an extra crop or switches to a higher moisture-demanding crop.

Common land

According to ASA, the increasing population and growing consumerism have taken a toll on the common land through encroachment and heavy demand on them. As a result, the common land has depleted in qualitative as well as quantitative terms, thus affecting the fuel as well as fodder supply. They realise that the condition of these common land needs to be addressed urgently if the rural economy is to be saved from bankruptcy. It believes that common land improvement is an important aspect of watershed development and therefore, ASA has given prime importance to it in the planning process. In ASA's area of operation, a substantial portion of the village common land has been brought under intensive treatment. Treatment measures include contour trenches, gully plugs, and plantation of trees and soil binding grass. Protection through social fencing is another important component of ASA's programme.

Water resources development

Gabion structures

According to ASA, gabion structures with a core concrete wall should serve as water harvesting structures. Considering the high seasonal variation in stream discharges, ASA believes that they are important means of meeting the irrigation needs of the *rabi* crop. They have persuaded the Self-Help Groups to construct a large number of gabion structures.

Check dams

ASA has been successful in rehabilitating several check dams, which were constructed by the government a number of years back but were lying dysfunctional because of neglect and lack of maintenance. With very little technical and financial inputs, the structures have been rehabilitated. ASA has succeeded in getting user groups formed for each of them. As a result of the rehabilitation of these structures, large areas have been brought under irrigation.

A little idea, a big difference

The check dam between Chulia and Karachat is dysfunctional ever since its inception. The state had the time and capital to make the dam but forgot to install the sluice gates. The villagers had no idea of how to make it functional, but they were industrious enough to remove the iron frames and bolts, the only parts of use to them. Thus, for four long years, the structure only added a touch of concrete to the earthy surroundings, and quietly flowed the Kadwal river.

It attracted ASA's attention when they entered the area. The villagers and the structure itself told the rest of the story. ASA discussed the case with the villagers first and an engineer from SWDF, Dahod later. On closer inspection, it was found that the installation of new frames might damage the structure. In this fix, Praful, the engineer, found an innovative way of using empty urea sacks to make up for the lack of frames. The villagers completed the rest.

And today the farmers of Chulia and Karachat take the *rabi* crop. Pray why didn't they do it before? But as the maxim goes, 'Better late than never'.

Ponds

According to ASA, ponds are the most common water harvesting structures in the region. This NGO has made pond construction and deepening of existing ponds, a mission. ASA has developed ponds in a large number of villages.

Wells

ASA has persuaded the Self-Help Groups to dig wells, and the activity is funded through the Self-Help Groups as loan to the member. These wells have contributed significantly towards increasing the area under irrigation.

Forestry

ASA has consistently stressed the importance of extension programmes. At their instance, the Self-Help Groups have installed a number of nurseries from which they have provided saplings of species like eucalyptus, bamboo, *subabool* and *khair*. Due to the abundant rootstock, natural regeneration with proper protection is capable of ensuring a good forest as well as fodder crop in the common lands. Artificial regeneration is therefore opted for only as a gap filling arrangement. ASA also coordinates with the Forest Department, and such efforts have been instrumental in developing several village forest management committees.

Agriculture programmes

ASA provides informal agricultural extension services to people during its work in the villages. In addition, it also provides advice on improved seed varieties of agricultural crops and vegetables and grafted horticultural saplings.

Agronomic practices

ASA believes that physical measures in watershed development can be fruitful only if they are supported by appropriate changes in the cultivation pattern and in farming technology. Their major emphasis has been on contour cultivation and seed priming.

Improved seed variety programme

At ASA's instance, improved crop seed trials have included composite varieties of maize, rice and gram. These higher yielding seeds are also drought resistant. ASA has also taken up trials and comparative demonstrations for these varieties for reliable impact assessment. The seeds are loaned to the farmers with an agreement of return at harvest time, and a seed bank is maintained for this purpose.

Compost

ASA has been instrumental in the construction of a number of Nadeps. This activity was earlier taken up on a pilot basis for the purpose of demonstration. Acute manure scarcity is a major factor accounting for the success of this programme.

Bio-gas plants

ASA believes that biogas plants are very feasible systems in the work area. Besides offering other benefits, this technology decreases work pressure on women. Two plants on a pilot basis have been constructed so far at ASA's instance. The programme is planned to be taken up on a larger scale in the following years. ASA's biogas programme is linked to the government programmes for subsidy benefits.

Bio-fertilisers

ASA has ensured that bio-fertilisers like azeto bacteria, rhizobium culture and PSB are introduced. The programme involves 105 farmers from three villages among whom 500 kgs of these fertilisers have been distributed.

Livestock development

ASA's activities in livestock development include pasture land management, cattle health programmes and breed improvement programmes. They believe that common lands are heavily grazed. Such pressure of undue grazing has left the pasture lands waste and barren. However, natural regeneration of fodder grasses is very high and if suitable protection is given, this would ensure abundant grass production. The efforts towards protection of pasture land have yielded results.

According to ASA, veterinary services are non-existent. In association with the state animal husbandry department, they have organised cattle health camps in the project villages. Realising that it is a crucial resource and needs to be emphasised, ASA foresees increased growth patterns in this programme.

Community capacity building

According to ASA, development is the process of change, preferably a guided change. Interaction among different cultures and learning from each other is a part of this process. Such interaction among various cultures and informed exchange of technology makes them mutually rich and more cohesive in the fast integrating world. They consider appropriate technological interventions and skill building as a part of this enrichment process. Such interventions not only prove to be sources of extra income but are also socially beneficial. For this purpose, ASA employs means like training and exposure visits.

They believe that training is a key element of the process. Since its inception, ASA has designed and conducted a variety of training programmes for its project participants and field level workers. Recognising that this is an enormous and continuous task, ASA has developed a moderately equipped training facility at the Bori Field Centre. The training modules, aids and materials are developed and continuously revised to incorporate the learnings. The training aids portray the local culture through Bhil language and visuals.

ASA organises exposure visits and result demonstrations on a regular basis for all the project partners and village level workers. Regular discussions with the project partners, in-house workshops and review meetings are other features of this programme.

The SHG members and other villagers have undergone various training programmes and exposure visits. The training areas are: SHG concept and management, leadership, watershed concept, planning and management, leadership, water harvesting structures, JFM, nursery raising, budding and grafting techniques, vegetable gardening, plantation technique and after care, gender sensitisation, etc. The impact of various skill training programmes and exposure visits is reflected in people's enhanced capacity to manage activities with low inputs from the project, especially for collective decision-making, planning and implementing activities, conflict resolution and adapting to new technological interventions.

In accordance with ASA's policy to develop local capacities, a group of village youths who are moderately educated are selected from project villages to work as volunteers. Since the knowledge base of the micro village situations is strong at this level, it is easier for these volunteers to take forward the implementation of project activities in a more effective manner. These volunteers have undergone intensive training and exposure visits on a variety of subjects, including a three months' apprenticeship at the beginning. The major training areas have been SHG—organisation, development and activation including SHG book keeping; watershed concept, planning and management; participatory methodologies of watershed planning; technical measures of watershed treatment including minor water harvesting structures; Joint Forest Management (JFM); nursery raising; budding and grafting techniques; vegetable gardening; plantations and aftercare techniques; Watershed Development Committee (WDC) book keeping; general office administration; awareness on government schemes; gender sensitisation, etc. The majority of them have visited MYRADA–Gulbarga project for an exposure to the operationalisation of SHGs to watershed development; SWDF, Dahod for water resources development and LI cooperatives management, Ralegaon Siddhi for participatory watershed management; PAHAL project and PEDO in Dungarpur and rainfed horticulture in Centre for Horticulture Research Station, Godhra. All these efforts have fostered confidence in the abilities of the participants to take forward the project implementation with very little managerial inputs from the organisation.

Social and economic programmes

Falia (hamlet) schools

ASA believes that because of Jhabua's peculiar geographical configuration in which the rural spread is so scattered, education should be taken to the *falia*s. ASA has always been a votary of opening as many *falia* schools as possible and decided to open schools in *falia*s where there were none.

She leads by example

She is dynamic and fearless. She walks with the confidence of a person in rhythm with nature. She is from the socially ostracised and economically poor harijan community, but Meeraben is a born leader.

It was because of Meeraben's initiative that the women of 'Harijan' phalia of 'Chulia' village were able to organise themselves in the form of a Self Help Group. She has again shown the way to the villagers by a rare entreprenurial effort and even rare business skills.

Meeraben purchased five goats out of loans taken from the Gramkosh. Before doing so, she not only surveyed the regional markets but also discussed the feasibility of the idea with people she considered worthy of knowledgeable advice on the subject. Within a period of five months, the number of goats had doubled. Another six months and Meeraben had returned the due amount to the Kosh. She took another loan of Rs 2000 and is today also the proud owner of a poultry business.

Her successful example has catalysed the process of attitude change among the villagers. They have also started withdrawing loans for purchasing agriculture inputs from the group funds.

Economic changes, we believe, are an influential precursor to social changes. And people like Meeraben are examples of emancipated human existence that lead to prosperity and egalitarianism. She shows the way, let's follow her.

ASA believes that *falia* schools should be managed by the village SHGs. The teachers should also belong to the same *falia*. The schools should be equipped with educational and recreation materials. ASA thinks that this activity will gain further momentum in the days to come, by networking with the state education department and the midday food supply programme. The number of schools will also increase.

Drinking water sources

According to ASA, drinking water, the most important of human needs, is the most difficult to access in the remote tribal villages. This NGO has, in some cases, tried to address this vital need by constructing hand pumps. The harijans are a socially ostracised community and have to face the maximum hardship for getting water.

Primary health programme

On ASA's entry to Bhil villages, it was confronted by an alarming health situation. In order to immediately address the situation, ASA organised a series of health camps. Soon after, a doctor was engaged. Initially the doctor regularly visited the villages and provided services and medicines, charging them 50 per cent of the medicine cost. Later, a dispensary was opened at ASA's Bori field office. Other aspects of the programme, which have gained emphasis include preventive measures (like immunisation programmes and provision of folic acid tablets to pregnant women etc.), health education and *dai* (midwife) trainings. These programmes of ASA are in liaison with the government health programmes.

Grain bank

According to ASA, income deficits have badly affected the concept of village self-sufficiency in the Bhil area. ASA's concept of grain banks is aimed at reducing the farmer's exploitation at the hands of the trader. Grain banks have been developed out of the Gramkosh funds. ASA's experience is that this programme has directly benefited the very poor section of the Bhil community, the additional merit being that it needs very limited amount of managerial inputs from this project.

Income generation

With ASA's encouragement, a number of villagers have started economic activities like poultry, goatry, shopkeeping and nurseries. This is in addition to their existing farming and livestock activities. ASA considers people's initiative for digging wells for irrigation purpose to be an important income generation venture. Dairy and lac production, the two activities that ASA plans for as major income-generating programme for the area, has potential for large-scale employment generation.

Gender in development

ASA believes that, like in most other cultures, the society in the Bhil area is male dominated. Women do more than their fair share of work, but their contribution is scarcely recognised. Their decision-making roles are also limited. According to this outfit, the process of development would have little meaning and would not be sustainable if equality between the two sexes is not maintained. This has to be ensured in the form of opportunities and decision-making roles. ASA has a two-fold strategy on this front. The immediate step is to reduce the drudgery of the women by improvement in fuel/fodder supplies and improved technological interventions like ball bearing systems for the household chakki (grinder), maize shellers, automatic pulleys and micro irrigation systems.

The second aspect of the strategy is to improve their status in the family and their role in decision-making. They play a primary role in the community-based organisations—more than two-thirds of the Self-Help Groups are women's SHGs. Norms ensure that there is only one representative from a family so that household integrity is maintained and the representative's powers do not get diffused. Considering the increasing role that these organisations are playing in rural life, there is bound to be a proportional rise in women's status in the family as well.

According to ASA, it has played a very meaningful role as the PIA for 25 villages in Udaigarh and Jobat blocks in the watershed project. It has been instrumental in creating awareness amongst the Bhil in these 25 villages on what is the root cause of the various problems confronting the Bhil and has infused confidence in them to find a solution to these problems. In the watershed project, it has organised the Bhil into viable groups which have acted as the agents of change in these villages. It has increased the capacity of the Bhil through several interventions. ASA has acted as a facilitator for bringing about noticeable changes through the watershed project.

Prayas

PRAYAS is an NGO working as the PIA for 10 villages in Bhabra Block of Jhabua district since January 1998.

Beliefs

PRAYAS believes:

- in co-existence of all life forms and in social, economical, political and gender equity among human beings;
- that sustainable changes in the rural resource management systems can only be brought in by the insiders and that they have the right to choose and decide;
- that indigenous systems technology and institutions should be prioritised as development options within the value framework of sustainability, equity and justice; and
- in the concept of self-reliant communities capable of control over the circumstances of their lives and livelihood.

The Path/Strategy

PRAYAS's strategy for realisation of its vision is as in the following:

1. Identifying, analysing and understanding problems of local rural systems and their linkages with natural resources management systems in partnership with the people.
2. Generating awareness among the people about the need for 'change' and kindle confidence in their ability to make it.
3. Mobilising and organising the community into institutions that can act as change agents at various levels.
4. Building of capacity of people through various interventions with emphasis on indigenous technology.

5. Facilitating development of sustainable natural resources management systems and livelihoods.

According to PRAYAS, there are two things wrong with the Bhil, namely degeneration of natural resources and low quality of human life.

Degeneration of natural resources

- Flawed macro level policies, infiltration of the dichotomies of mainstream civilization and their incoherence in the context of an inherently different society and increasing rural population are among the factors that have not only led to irreparable damages to the natural resources stock but also continue to threaten whatever remains, simply by their huge demands. The fact that land management systems have not been able to cope with the induced changes in local rural conditions has only exacerbated the destruction of the natural resources. The condition is going from bad to worse. A land which, merely 50 years ago, was covered with thick forests is today known for its 'barren & rocky' satellite images. A large part of this erstwhile forest land is being ostensibly cultivated. However, unscientific cultivation of such lands, specially those with higher gradients, can scarcely qualify as 'agricultural land', for they yield little and are most often infested with roots and shoots of various species.

Low quality of human life

- Factors like destruction of forests, increasing population, lack of a philosophy of maximising production in the very nature of the prevalent traditional agricultural systems and its ironic co-existence with the newly acquired 'hoarding' values have got interwoven to create shortages of food, real or artificial. Most of the agricultural land is rainfed and faces frequent droughts. This risk factor also forces the farmers to opt for drought resistant but low yielding local crop varieties.
- For the last few decades, food shortage has been the primary factor responsible for 'seasonal migration', a phenomenon that has become a feature of the Bhil region wherein 60 to 70 per cent of the

rural population migrates to the regional cities for a period of three to four months.

- The present pattern of utilisation of the village resources is unable to sufficiently meet the needs/demands of its inhabitants. The average village economy is a deficit one. This is because of the almost complete dependence on 'agriculture' in which practices remain primitive. The economic deficit is compounded by factors like pressures for high social expenditures and the desire for conspicuous consumption, features acquired in the close past. As a result, middlemen and moneylenders rule the day and interest rates are as high as 150 per cent p.a. Loans are paid back only to be borrowed again, with interests deducted beforehand. These factors most often land the common people in a vicious cycle of exploitation and recurring indebtedness.

- Due to cultural and social pressures of the outside world, the infiltration of local politics and increasing role of money in the tribal economy, the ancient life values of the rural society like collective living, equality, frugality and co-existence have been deeply affected. The trend towards hoarding of private wealth and desire for conspicuous consumption has sown seeds of enmity while co-existence as the basis of society has been broken down by the vagaries of the modern economic conditions, administrative mechanisms and political environment.

- The indigenous breeds of cattle are often undernourished and qualitatively poor in terms of milk produce and draught quality. Because of lack of proper veterinary care, the general health of the livestock also suffers.

- The average rural woman leads a deprived life. She is not only marginalised within the household but also suffers because of the poor economic and social condition of the 'household' itself. The women are also socially marginalised and have no role in the community decision-making processes.

- The literacy rates are low (approximately 5-6 per cent of the men and 2-3 per cent of the women are literate). With poor education facilities and their lack of sensitisation to the needs, the status shows no distinct signs of improvement. The health status also suffers because of the inability of the traditional medicine systems to cope with the influx of modern diseases and the lack of access to modern medicine systems.

- The villagers are not organised on development issues and are unable to negotiate with the government delivery systems.

According to PRAYAS, the empowerment of the Bhil can be done with the following steps.

Community organisation

Self-Help Groups (SHGs)

According to PRAYAS, SHGs are institutions of social reconstruction that are supposed to outlast specific projects and ensure the development of the community at its terms in the future. Being the primary change agents, they are the lifeline of the programme and occupy a position of pride among the CBOs. The savings and credit activity on account of the multiple benefits it provides forms the ideal adhesive for these institutions.

The membership of the SHGs that PRAYAS has helped organise varies from 10 to 20. Social comfort among members of the same SHG is necessary for effective and sustainable functioning of these institutions. Kinship relations often define this factor. The SHGs maintain their respective funds, consisting of regular periodic contributions from the group members as 'savings'. The group unanimously decides the individual contribution amount. It is necessary that this amount should be such that the poorest member of the group can afford. The contributions are made every fortnight in a meeting, which is the forum for discussions and decisions on various issues of village development. Village volunteers, selected by the groups, facilitate the meeting. The group representatives deposit the savings in a nearby bank, every month. This savings account operates as a joint account of two group representatives.

According to PRAYAS, women have been more forthcoming than men because of the multifarious benefits that savings offer. On the one hand, it secures and improves their family condition, on the other, 'personal savings on thy own name' is a big boost to the average adivasi woman who has little ownership rights over resources. The impact of 'savings' is also evident in a number of social and cultural manifestations. Women's groups, therefore, far outnumber men's groups. The respective SHGs fund the credit needs of the members either from the SHG fund or by networking with formal financial institutions and government schemes. Credit is provided for entrepreneurs as well as contingency purposes. Micro credit operations, especially among women, enable better investment in the household and farming sectors and for entrepreneurial purposes, and are necessary for making sustainable improvements in the quality of life of the people. The credits are provided at a rate of 24 per cent p.a. to the members.

Watershed development project

In the watershed villages where PRAYAS is the PIA, the work is planned and monitored by the Watershed Development Committees (WDC), which are representative bodies of the whole population of the respective villages. The president heads the WDC while the secretary is responsible for its activities, general administration and accounting. While the president's position is honorary, the secretary is a paid employee of the WDC. The Self-Help Groups because of their kinship, neighbourly and friendly ties, form efficient units of implementation in the field. Members of a SHG are usually from the same hamlet. It, therefore, becomes possible for two–three neighbouring SHGs to come together in the form of activity-based groups (user groups) to implement activities in specific village parts. The cluster of SHGs selects a volunteer to facilitate the routine implementation work. The WDC pays the people in terms of units of work done. In the accounting process, the WDC secretary supports the volunteer. The president, secretary and the volunteer are present during payments. The WDC maintains three major accounts; namely the project account, the Gramkosh and the Vikas Khata. The project account is solely for the purpose of receiving funds from the Zilla Panchayat while the Gramkosh and Vikas Khata are village development funds maintained by contributions from the villagers.

According to PRAYAS, the Gramkosh is used to fund various village development activities. It is also a source of refinancing for the SHG funds. It is a village level fund maintained by villagers' contributions out of their labour earnings for project work. The people contribute 45 per cent of their earnings from watershed work to the Gramkosh. The contributions have an important role to play in the sustainability of the programme as they not only serve as the means of people's identification and participation in the work but also positively influence the savings–investment cycle and lead to rural reconstruction and development.

In the watershed villages looked after by PRAYAS, the various Gramkosh together have a balance of more than Rs 10 lakh. In addition, more than Rs 75,000 are under circulation as refinance for SHGs. The Gramkosh have funded, either partially or fully, a number of village development activities like construction of village roads, electrification of community buildings, etc. The Vikas Khata is a futuristic fund collected for the maintenance of the measures after the completion of the project. The people contribute 5 per cent of their earnings from project related work in this account. It has a balance of about Rs 2 lakh.

According to PRAYAS, the watershed development work is based on an action plan, which contains baseline information in addition to specific activity targets. While most of the required information is collected through participatory rural exercises, the rest is obtained from secondary sources in the district and block administration. The plan, in turn, is prepared with the active participation of the people.

Soil and moisture conservation

In the watershed villages where PRAYAS works, the physical work in land improvement is based on the ridge to valley approach. Common lands as well as private lands are considered for treatment. These treatment measures include contour trenches, contour bunds (earthen and stone), gully plugs and loose boulder check dams. In the undulating terrain of the working area, processes like contour trenches, gully plugs and loose boulder check dams are very useful for the conservation of soil and moisture in the fallow lands. Complemented by plantation of soil binding grasses on the bunds and tree species in the rest of the land, these measures are of immense help in restoration of the village ecosystems. The ecological restoration, in turn, has a direct effect on the people's livelihoods through the agricultural and socio-economic impact it creates.

According to PRAYAS, cultivation of land with high gradients without proper measures for soil and moisture conservation is one of the most threatening problems that the village land faces. This is because it is rendered 'devoid' of as precious a resource as 'soil'. Due to the increasing demands, while cultivation of such lands is not expected to stop, the way out is to minimise the damage. This would include treatment of such land with conservation measures in the form of stone bunds, earthen bunds and gully plugs.

So near yet so far

The average adivasi village is not untouched by the anomalies of the Indian nation at large. Differences in economic and social status often mean different priorities and perceptions. The extent to which the privileged segments understand and feel the needs and wants of the unprivileged segments is 'limited' by many a constraint.

It was in October 1998. The physical work in the watershed project in village Kakadbari was to commence. For this purpose, the PRAYAS team initiated discussions among the WDC members and also discussed the issue with the Sarpanch, Patel or Patil and some other influential members of the village, for it remains a fact that these people, if not taken into confidence, can become real stumbling blocks. These sections showed little initiative instead of saying that this being an agriculturally demanding season, the people were busy in agricultural work. However, literally the whole village could be seen migrating with people moving out in groups of five or six with their food stocks and other daily requirements.

Finding that little help was coming from this quarter, the team decided to approach the women's SHGs. They had a contrary opinion that the lack of jobs in the village was fast pushing the people into migration. The groups represented the opinion of the common people and were therefore more dependable. In a joint meeting of the women's groups, it was unanimously decided that work be begun forthright, for it would not only provide an income source to the people within the village but would also improve the condition of the village land and resources.

Madiben of Simada Mahila Samuh, an enterprising woman in her forties, took the lead in the process. Soon, even those who had migrated were called back and the planned physical activities for the year were soon over.

Maliben provided similar leadership in Sanda village. She is an important resource person for any work in this village. These incidents make it clear that where ensuring 'equity' is concerned, women's leadership can be very effective.

Development of farming

In the villages where PRAYAS works, agriculture is the primary occupation within the villages. It is supported by other farm-based activities like livestock, rearing poultry, handicrafts, etc. However, due to inefficient management, these activities are unable to meet the livelihood needs of the people. Only less than 20 per cent of the households harvest enough to last them throughout the year. Livestock rearing is equally unprofitable, with an average cow yielding less than a litre of milk a day during its peak productivity period. PRAYAS has worked towards sustainable farming systems that would lead the villages towards self-sufficiency in terms of their livelihood needs. 'Watershed Development' for improving the quality as well as the utilisation pattern of land provides the basic foundation to work towards this end. This is supported by introduction of higher yielding seed varieties and other interventions in in-farm as well as off-farm sectors.

Agricultural interventions

Participatory introduction of new crop varieties

According to PRAYAS, though the traditional crop varieties best suit the climatic conditions of the Jhabua belt, they have been unable to cope with the demands of increasing population and the prevalent market economics. PRAYAS has therefore tried to introduce the people to new seeds which can take care of the emerging needs along with the local crop breeds which the farmer is likely to continue with because of the long 'trust-based' relationship that he/she has with them. PRAYAS has done the following:

- Volunteer farmers have been provided with a number of seed varieties to be grown in rows in the same field along with the existing varieties. This has helped in giving a true idea of the output of the various introduced varieties as compared to the existing varieties in exactly the same conditions. This has also helped in site-specific evaluation of the success of each of these varieties.
- PRAYAS has helped several farmers from a number of SHGs to establish seed production plots. For this enterprise, they were provided with foundation seeds funded by the watershed project, as loans to be repaid after harvesting. The purpose of the seed production plots is to serve as sources for the seed banks, to be established in every village. The farmers have individually stored a part of the seed produce with them for their own needs and also for sale to the villagers while the rest was sold (10 per cent to 15 per cent more than the market price) to the agricultural department, which later sold them as 'certified seeds' after sorting.

Demonstration of new irrigation systems

According to PRAYAS, on account of scarcity of water and the undulating nature of the terrain, flood irrigation is not the ideal way of irrigating fields in this land. Irrigation, however, becomes necessary considering the pressure to produce more in the current conditions. If innovative ways to cut down the costs can be found, modern scientific irrigation systems can play an important role in these conditions. Nobody is a greater expert in finding such site-specific ways than the average Indian farmer. The introduction of such systems by PRAYAS, namely the drip and the sprinkler irrigation systems is a testimony to its trust in the ability of the Bhil farmer, who, it believes, will make them successful inspite of the projected cost of such systems. PRAYAS feels that the success of these experiments could well change the look of these fields.

Forestry interventions

According to PRAYAS, tree plantation and regeneration of the forests is the key to the reconstruction of the average Bhil village. This is so because of its unique geographical, climatic and demographic characteristics. Since all forests in the work area of PRAYAS stand destroyed, plantation is the only way left. The outfits' attempt has been to facilitate maximisation of plantations. Most of these plantations have been private efforts, facilitated by the group institutions.

Fisheries

At PRAYAS' instance, two user groups of 10 members each from the villages of Kheriamali and Kasat have introduced fish of the katla breed in the ponds in their respective villages. The user group membership was open and the activity was initiated with the permission of the Gram Sabha. 20 per cent of the net profit from the activity was deposited in the Gramkosh.

Pasture development

Pastures have been developed in 11.6 hectares of revenue wasteland in Chapri village. This land had been encroached by a few neighbouring families of the village and was being used for livestock grazing. Due to indiscriminate grazing, the land offered little in terms of fodder to the livestock there.

(Box Continued)

> PRAYAS initiated discussions with the people to allow treatment of this land so that it could not only provide fodder for stall-feeding, but could also be planted with tree species. It was difficult initially but after extensive deliberations, the farmers agreed to allow treatment of a part of the land though they did not agree to refrain from grazing. The encouraging results motivated them to allow treatment of the rest of the land also. This was followed by their initiative in stopping grazing as well, when the villagers reached an understanding that the livestock belonging to the families in question could graze in the rest of the village common lands.
>
> The land was planted with grass breeds, namely, Stylo hamata and Dinanath, and more than 800 plants of tree species like Eucalyptus, bamboo etc. In the very first year, about 9,000 pulas (bundles) of grass were harvested from this pasture land. The villagers paid for the grass at 50 per cent of its market price and deposited it in the Gramkosh. The success of this experiment has been an eye opener for PRAYAS. It proves that winning the trust and confidence of the villagers was the first step towards getting them interested in the reconstruction of their village habitats.

Livestock development

According to PRAYAS, the livestock quality is poor, as are the fodder, health and general living conditions of the livestock. Minor injuries, infertility and diseases like Foot and Mouth disease (FMD), Haemorrhagic Septicaemia (HS) and Black Quarter (BQ) are common problems. However, livestock rearing remains one of the major occupations of the Bhil community. PRAYAS feels that it is therefore necessary to improve the general condition of this occupation. In addition to pastureland development and plantation of fodder tree species, PRAYAS has also taken up conducting veterinary health camps.

Income-generating activities

Twelve multi-purpose veterinary health camps in partnership with the Bhabra Tehsil veterinary hospital have been organised in 10 villages. The camps served the purpose of building awareness on crucial issues of livestock care and curative and preventive treatments. A total of 2,850 livestock were treated. The farmers of Dunglawani, have borrowed from the Gramkosh funds and have purchased an he-buffalo. Interested farmers of neighbouring villages use this facility. They are expected to pay an amount of Rs 150 per service.

Income-generating schemes

At PRAYAS' instance, individual members of the SHGs have ventured into a number of income-generating activities. Most of these activities are built on indigenous resources and technology. The following activities give an idea of the income-generating schemes taken up.

- 15 women have borrowed from their respective group funds and DWACRA for establishing brick kilns. The kilns use traditional local technology. The enterprise is expected to be successful, considering the easy availability of raw material, sufficient demand and good quality of the bricks made by these methods.
- Five women have borrowed loans for starting poultry business. They have purchased a hundred chicks each. A net profit of Rs 30 to 35 is expected from each chick. Considering poultry is an established tradition and an important livelihood source in the Bhil community, this entrepreneurship has high potential.
- 10 women from a group have borrowed loans for investment in the craft of rope making from sun hemp. Besides using these ropes for domestic purposes, the cots made of them will be sold within the village and in the local markets.
- One member of a group has purchased a thresher for hiring purposes.

Federation of self-help groups

At PRAYAS' instance, a federation of all the Self-Help Groups in the 10 watershed villages has been set up. Mahila Mahasamiti is a federation of the women's Self-Help Groups. The general body of the Mahasamiti consists of the members of all SHGs in the 10 project villages. Every village is represented by a minimum of two SHG members in the 21 member working body. The federation supports individual SHGs in terms of linkages with the government programmes, bulk purchases and supply, providing services, policy advocacy, etc.

Grain bank

According to PRAYAS, food insecurity is a major problem in the Bhil villages. In order to alleviate the problem of food insecurity, which almost four out of every five adivasis living in the region face, grain banks have been set up. Earlier, the people had to depend on the local businessmen who exploited their position by charging interest rates as hefty as 100 per cent per season; now the people depend on these grain banks. The grain banks are initially established by taking advantage of the funds of the watershed project. But they are now self-sufficient as the beneficiaries deposit 25 per cent more than their borrowed quantities after the harvest. The concept has, however, been changing with *falia*s and SHGs having their own grain banks, especially in the view of the scattered household location in typical Bhil village. These grain banks have the potential to ensure that nobody goes hungry in this drought-prone land.

Women's empowerment

According to PRAYAS, in terms of working hours, women on an average work about 15 to 20 per cent more than men in the Bhil area. They not only look after the domestic work but also share farm work, which is not very differentiated. Women as a group consume far less than their fair share of the household resources but lead a life which is more difficult than men. Social equations and resource distribution criteria also work towards relegating them to a secondary position in society/ keeping them in an inferior place in the society.

PRAYAS has adopted a multi-faceted strategy for sustainable improvement in the quality of life of the average adivasi woman. While on the one hand, efforts are made to make sustainable improvements in the status of the household and to maximise women's involvement in the process, on the other, reducing the drudgery of the daily life of the average adivasi woman is a priority issue. Most interventions encouraged by the outfit aim at improving the household condition, which is expected to make a difference to the women not only by the trickling down of the benefits but also by their focus on women's involvement. This has been specially so in activities like savings, income-generating activities, etc. Most credits into the villages are channelised through women's SHGs. At PRAYAS' instance, the large number of forums and formal institutions created, provide the women opportunities to not only increase their confidence and articulation skills, but also to voice their concerns and feelings. Fuel wood collection and cooking food account for almost one-third of women's daily time. To reduce the average adivasi woman's drudgery in this activity, large-scale plantation of fuel wood species has been conducted and gobar gas plants have been constructed. The plants were funded jointly by Khadi Village Industries Commission (KVIC) and the watershed project. The beneficiaries are expected to contribute Rs 1,000 into the Gramkosh. Since the Bhil women also take care of field preparation work, PRAYAS has actively advocated for the establishments of NADEPs to reduce women's drudgery in field preparation.

PRAYAS believes that for an improvement in women's social status in these societies, the preservation of their traditional institutions, practices and social elements is essential. As the traditional institutions, practices and social elements have deteriorated with the onset of modernity, the Bhil women who enjoyed a far more equal position in the past now find themselves in a secondary position in Bhil society.

Capacity building

Capacity building is an integral component of PRAYAS' strategy for realising its vision. Interventions in capacity building include training and workshops on specific issues and exposure visits. Confidence building and skill improvement are the major objectives of these interventions, which have been arranged for the community as well as for the PRAYAS team members.

The outfit believes that its contributions as the PIA of 10 villages in watershed project have borne fruit. The watershed activities have led to an increase in the double cropping area as well as the *rabi* irrigation potential in these villages. There is an increase in cropping intensity. All the community organisations established as a part of the watershed project are working well and have contributed to the empowerment of the Bhil. Self-Help Groups have been particularly active in mobilising savings and have fulfilled the credit needs of the Bhil villages. The quality of life of the Bhil women has improved substantially and they now have a voice in the society.

Assessment of the role of the NGOs

What came across from interactions with the NGOs is that they have been able to win the trust of the Bhil community and establish necessary rapport with them. This is not an easy task in the Jhabua setting, considering the fact that the Bhil are wary of initiatives emanating from outside agencies. They were in a position to win the confidence of the Bhil and get accepted by them by taking up entry point works which ranged from conducting health camps to construction of bus stands.

One important thing that came across from interactions with the NGOs is that they have been able to create awareness among the Bhil as to the cause of the various problems confronting them. They have succeeded in persuading the Bhil to identify, analyse and understand the problems of a local rural system like in Jhabua district and to appreciate the linkages of these problems with the natural resources management system. They have been able to make the Bhil see how their very own acts of destroying the forests and cultivating the common property resources have been responsible for the degradation of the ecological system and the resultant drought conditions year after year.

Another important aspect is that the NGOs have succeeded in sensitising the Bhil to their self-worth, self-respect, and rights and responsibilities as legitimate members of the society. They have also succeeded in generating awareness amongst the Bhil as to the need for change and giving them the necessary confidence in their ability to be the instrument of change themselves. This is a major achievement, considering the fact that the Bhil are usually diffident about being in charge of their own lives and livelihoods, after centuries of disempowerment and exploitation.

The NGOs have also succeeded in mobilising and organising the Bhil into groups and community institutions which can act as change agents at various levels. They have been organised into Self-Help Groups, user groups, watershed associations and Watershed Development Committees. By motivating the Bhil to form themselves into functioning and viable community organisations, the NGOs have been able to bond social capital amongst the Bhil. This has facilitated collective action, which has enabled the Bhil to increase their access to resources and opportunities, and participate in the process of governance.

These NGOs have also built capacity in the Bhil through various interventions such as training, exposure visit, awareness-raising, etc. They have also been instrumental in enhancing assets and capabilities of the Bhil, translating and interpreting information to them, and providing them with technical services. The NGOs have also functioned as important links between the Bhil and the government departments, and in that sense, they have played critical roles in linking social capital.

There are two aspects that have been a common refrain while discussing the association of NGOs in rural development projects. First, the NGOs have, in many cases, not reflected the interests of the rural poor in their activities, and they have been partial to their own interest (The World Bank 2000). Second, the association of the NGOs has inhibited the creation of enduring institutions at the village level. A common weakness of NGO-led development programmes has been that they have tended to foster a culture of dependence, and community processes set up by them during the implementation of a project have collapsed soon after their withdrawal (Shah 2006).

How far are these true of the functioning of the NGOs in the Jhabua watershed programme? Regarding the NGOs not reflecting the interests of the rural poor, it came across very clearly from the inter-actions with the NGOs in Jhabua district that they have reflected the interests of the Bhil. This can perhaps be explained by the fact that these NGOs are essentially manned by development professionals who have been working in the Bhil area for a number of years. Their first-hand grassroots experience has given an integral sense of identity with Bhil interests.

As far as the second concern that the NGOs tend to foster a culture of dependence and community processes collapse with their exit, the NGOs were quite emphatic that they were there to act as facilitators to help achieve the development needs identified by the Bhil themselves, and the Bhil community institutions would survive their withdrawal from the scene. But this is something that only the Bhil can tell.

8

People's Perceptions

The watershed project in Jhabua was planned as a participatory project. It was envisioned that the Bhil, after being suitably educated, would be in a position to understand and appreciate the problems of the watershed, and they would prioritise problems, identify solutions and prepare action plans. They will organise themselves into groups and community institutions such as the Self-Help Groups, user groups, watershed associations and watershed development committees, which will implement the project and manage the works. It was also envisioned that these community institutions managed by the Bhil would enable them to establish their role in the decision-making process in village development, and would give them a voice in the processes that shape their life and livelihoods. Has that happened? The following assessment is based on interactions with the Bhil who participated in the watershed project in Kolyabada, Choti Malpur, Badi Malpur, Sanda, Mankankvi and Badi Dhami villages of Jhabua district.

Kolyabada village

At the time of the visit, Kaliya, the president of the watershed development committee of Kolyabada village, was present and so was Shri Rama Bhai, a member of the Watershed Development Committee. Vesti, Dhuri, Rambi, Dadmi, Thooti and Kundlia, all members of the Self-Help Groups in Kolyabada village, were also there. They provided the following information about the watershed project.

Watershed associations, Watershed Development Committees, user groups and Self-Help Groups have been formed for Kolyabada village. All households in the Kolyabada watershed are members of the Watershed Association which is the supreme body in all matters relating to watershed activities. Most households including the women and poor did actively participate in the planning process of the watershed and formulation of the action plan. They have been given exposure to the concept of watershed, and a number of villagers were taken on exposure visits to places where watershed projects have been successful. They had also received skill training, both technical and managerial. Most households in this village had been organised into Self-Help Groups and user groups.

The members of the Watershed Development Committee have actively participated in the formulation of the action plan for Kolayabada village and also its implementation. The Committee has built check dams, ponds and other soil conservation works on the hill-slopes and has planted grass there.

'How has the watershed development committee worked?' I asked
'Let me explain', said Kaliya, the president of the watershed development committee. 'It's the committee of all the Self-Help Groups in the village and of the user groups. We sit down and plan the works. That work on the hill. The check dam. The ponds. Then, we got the cement, the steel and everything else. Do you know how much we spent on these works? Something like Rs 6.5 lakh'.
'Why, that's fantastic!' I said. 'What about doing accounts for the works you did?
It was Shri Ram Bhai who spoke now. 'We do that, alright. But let me tell you something. Everyone knows how much has been spent. Because we have given our labour. It's our own work, isn't it?'
'What about the work on the hill slope?' I asked.
'That's our work, too', Shri Ram Bhai said. 'But it belongs to the entire village. We grow grass there for the village. We don't allow anybody to graze cattle there'.
'What happens to the grass?' I asked.
'Part of it we share and part of it we sell', Kaliya said. 'Do you know that before we started growing grass on that hill slope, we had practically no fodder in the village? We had to go and buy grass from Bori to feed our cattle. Now we sell fodder to other villages'.

According to the villagers of Kolyabada, they have benefited from the watershed works which they themselves have planned and implemented. The works have prevented soil erosion. The availability of water has increased, and there is an increase in the groundwater resulting in higher levels of water in the irrigation wells. This has enabled the villagers to take on a *rabi* crop. The villagers grow grass now on the hill slopes; this meets their fodder requirements.

'You formed these groups and took up all these works', I said. 'Now, let me ask what have you got from them?'

Kaliya rubbed his chin, as if lost in thought. 'What have we got? A lot, really. You see, the soil does not run down that hill slope any more. Did you see the hill when you came to our village? Did you see how green it is now?'

I nodded.

Kaliya brightened. 'Do you know that it's the only green hill on this side of Bori? What a lot of grass! But that is not the only thing. With all these dams and tanks we've built, we have much more water now'.

Kaliya took a deep breath and closed his eyes. He had the satisfied look of a man who has achieved a mission.

'Why don't you come and look at our wells?' Kaliya suggested. "There's so much water now. More water now than we ever had. And what about the water for our fields? We can now take a *rabi* crop. Because we know that we can get water when we want it. We never took a *rabi* crop before in Kolyabada village. Oh, no!'

The villagers of Kolyabada felt that the watershed project had made a big difference to their lives. Before the project was started, they were ignorant about proper agricultural practices, with the result that they got very little yield from their lands. They had no idea about soil and water conservation. They had cut down the trees and cultivated the forest land, and in the process, had caused serious degradation to the environment. They learnt about their mistakes from exposure visits and have put that knowledge to productive use.

The villagers informed that all the community organisations formed in the village under the watershed project like the Watershed Development Committee, the Self-Help Groups, and the user groups have developed norms for working and use them while deliberating in or with these bodies. The Watershed Development Committee in Kolyabada has developed its working norms and has got them approved by the watershed association. The Watershed Development Committee has created transparent systems of work approval and financial transactions. All the details of the financial transactions are read out in the monthly meeting of the watershed association.

Regarding the accountability of the Watershed Development Committee and watershed association to the Gram Sabha of the Village Panchayat, the villagers said that the Gram Sabha was not a representative body of the village and they were not invited to the meetings of the Gram Sabha. They were of the view that the government works taken up in the village were planned and implemented unilaterally by the village panchayat without their participation.

The villagers said that the Self-Help Groups in Kolyabada were the nucleus of all development activities in the village including planning, implementation and review. Savings and credit were the core activity of the Self-Help Groups. The groups followed the principle of single representation for each household.

There are also user groups in Kolyabada which were given the task of managing specific activities. The majority of these user groups looked after the waterbodies created by the watershed project. The villagers had formed a Gramkosh. The Gramkosh is the fund formed from the contributions of the villagers out of the payment they got for watershed activities. Money obtained from other sources such as selling of grass grown in common property resources of the village was also credited to the Gramkosh. The funds of the Gramkosh were used for refinancing the Self-Help Groups, running the grain bank, meeting the expenses of the primary school and building community structures.

The Bhil in Kolyabada do not go to the Badwa, the witch doctor, nor do they go to the moneylender. All their credit needs, both consumption and productive, are met by their own Self-Help Groups. These groups have been able to mobilise a substantial amount of savings from the members themselves, have availed of loaning facilities from the financial institutions, DWACRA, SJSY and NABARD schemes. The groups have also taken money from the Gramkosh.

'Tell me about the Self-Help Groups', I asked.

Vesti brightened. 'Well, that's our own group. We're members. We all sit down and decide about things'.

'How often do you meet?' I asked.

'We meet once in a fortnight', Vesti explained. 'Either during the day or at night'.

'How many Self-Help Groups do you have in the village?'

'We have three of them', Vesti said. 'One is for women and the other two are for men. You see, each family has only one member in the group. We don't allow more than one member from a family'.

'What do you do in the group?'

'We save a fixed amount each month and deposit it', Vesti continued. 'We use this money for giving loans to members who need it. We also buy fertiliser and seed with this money'.

'How do you decide who should get a loan?'

Vesti gestured to Dhuri. 'Why don't you tell him, Dhuri? You are the one always talking about loans, isn't it?'

Dhuri or Bhuri spoke up. 'Why are you always after me? All right, I'll tell. All the members of the group sit down and decide. The first time we met, we decided how much money a member should give every month. When we had enough money, we decided to give loans. The members ask for loans. Even for marriages and funerals. We also decide what interest they should pay'.

'What if the loanee doesn't pay back?' I asked.

Bhuri made a face. 'Why shouldn't they? They all pay back. If they don't, we know what to do. We go and sit in front of the loanee's house and won't leave till the loan is paid back'.

'Don't your husbands object to your contributing to the group every month?' I asked.

It was Vesti who replied with a grin. 'Oh, they don't. If they do, we'll ask them to go to hell. Things have changed. They don't object now. Why should they? It's all for the good, isn't it?'

The villagers have also established a *falia* school in Kolyabada and appointed a teacher. Now all the children in this village go to the *falia* School. Before this school was established, the children had to negotiate long distances to go to school and as a result, many children had dropped out of school. This school is established and managed by the Self-Help Group. The funds for this school are provided out of the money in the Gramkosh. Because of easy credit facilities available from the *Bairani Kuldi*, a number of villagers have started income generating activities like poultry, goatry, running of small shops and raising nurseries. While some Bhil have also started dairy and lac production activities by taking loans from the Self-Help Groups, some others have taken loans for digging wells for irrigation purposes. In general, there is plenty of work available in the village itself, and as a result, migration has come down considerably.

The effective functioning of the community organisations has led to considerable empowerment of the Bhil women in Kolyabada. The women have successfully managed the Self-Help Groups, and have participated meaningfully in the working of other community organisations. Apart from economic empowerment, they have a say in the decision-making process that affects their lives. The improved position of fuel and fodder supplies and drinking water has reduced the drudgery faced by the Bhil women in Kolyabada. In addition, the introduction of improved technological innovations such as ball-bearing systems for the household grinder has helped reduce the drudgery of women. Asked what they will do after ASA, the NGO that is the PIA for Kolyabada village, withdraws, the villagers were ambivalent. It was obvious that the villagers would have difficulty in managing things.

Choti Malpur and Badi Malpur village

At the time of interaction with villagers of Choti Malpur and Badi Malpur villages, Karam Singh, the president of the watershed development committee of Badi Malpur village, Ratan Bhabhar, the president of the watershed development committee of Choti Malpur, Besti Behn, and Bhuri Behn of the Self-Help Groups of Badi Malpur were present. There were also a large number of other Bhil men and women from these two villages.

The villagers stated that a large number of watershed works had been taken up in both the villages, which included field bundings, both earthern and stone, in private lands. Binding grasses like Stylo hamata had been planted over the bunds. Common lands had been given treatment that have included contour trenches, gully plugging and plantation of trees and soil binding grasses. Protection of the common land through social fencing had also been done. The common lands in both the villages had been taken over by the watershed development committees for operation and maintenance.

The villagers stated that as a result of these measures, at least half the total rainfall and much of the available run-off has been conserved. Almost all the private land in both the villages has been treated. According to the villagers, the double cropping area in both the villages has increased. So has the *rabi* irrigation potential. As a result of the watershed works, water is also now available to sustain the *kharif* crop, in case there are drought conditions. There is an increase in cropping intensity. The villagers stated that they have also introduced new crop varieties in consultation and with encouragement from PRAYAS, the NGO that is the PIA for the watershed project for the two villages.

'Tell me about the watershed development committee', I asked Ratan. He rubbed his temple. 'Well, our committee is not very different from the groups. Only that it's a bigger group. It has members from the Self-Help Groups and user groups. Some members from the Gram

Panchayat. And a person from PRAYAS. It has a president and a secretary. The committee pays the Bhil wages for the watershed work they do'.

'What kind of watershed works has your committee done?' I asked him.

'Oh, all kinds. Land has been treated. Both common land and private land'.

'Who decides what land should be taken up for treatment?'

He waved his hand around. 'All of us. It's there in the action plan. Do you know who prepared the action plan? All of us'.

'What kind of treatment work have you done?' I asked him.

'We have done trenching. Bunding in stones and earth. Check dams with boulders. We have also done plugging. What do you call it, Dilip Bhai?'

'Gully plugging', Dilip Dave said.

'Yes, that's what it is', Ratan Bhabhar continued. 'We have put grass on the bunds. We have also planted trees'.

'Have these things helped?' I asked him.

Ratan stuck a finger in the air. 'Oh, yes! You see, now we have enough fodder in our villages. We even sell fodder that we don't use. There is more water now. And enough work here'.

The villagers have also taken up afforestation in a big way. They have raised nurseries, with seeds and poly packs being provided from the funds of the watershed project. *Khair*, eucalyptus, bamboo, papaya and mulberry are the major saplings being raised in the nursery. The nursery in Badi Malpur village is right on the roadside and is ideally equipped to fulfil the plantation needs of the neighbouring villages. According to the villagers, it is functioning as a central nursery because of its well-connected location and easy availability of inputs.

The villagers have taken up a number of income-generating schemes, by availing of loans from the Self-Help Groups. These schemes include brick kilns, poultry, rope-making, etc. The brick kilns have been successful as income-generating ventures because of easy availability of raw material, sufficient demand and good quality of bricks being made using traditional local technology.

Watershed Development Committees, user groups and Self-Help Groups have been formed for both these villages. These organisations are working well. All the Bhil women have been organised into Self-Help Groups. They were actively involved in planning, implementing and monitoring of watershed works. The Self-Help Groups in these two villages have generated savings from the members by way of regular monthly contributions and have mobilised and rotated the money as credit amongst their members. They have also mobilised external resources for meeting the larger productive needs of the members.

The Bhil women stated that the institutions established in the watershed project such as the Self-Help Groups, user groups and Watershed Development Committees have created opportunities for women to express their views. In the process, they have been empowered to establish their role in the decision-making process in the village. According to the Bhil women of these two villages, institutions such as the Self-Help Groups, user groups and watershed development committees have been sensitive to the causes and interests of the women and poor in the Bhil community.

'Now, tell me the truth', I asked the Bhil women. 'Have you people benefited from all your group work?'

'Yes', Bhuri *Behn* said, 'Before the group started, we had no money. If we needed money, my husband went to the moneylender and borrowed money. And to pay the moneylender back, we had to migrate for six months a year. You see, in the past, our husbands never told us anything. They never consulted us. That's because we are women. Because we didn't have money. But now we have money of our own. You see, that makes us somebody. Now our husbands know that we count for something'.

'That's true', Besti *Behn* intervened to say. 'Because we save money in the group, we buy fertilisers in time so that the crop is good and we don't have to suffer. And nobody has to go to the moneylender. Isn't that a good thing?'

'Yes', I agreed.

Besti *Behn*'s eyes twinkled. 'Let me tell you about the *Kuldi*. We have this *Kuldi*. Do you know what that means? It means a kind of pot in our Bhil language. We keep about Rs 500 in reserve in *Kuldi* after all the contributions have been deposited in the bank. Suppose, now, somebody's child were to fall sick. She'll go *the Kuldiwalli*'s house, take the money from the *Kuldi* and rush her child to the hospital'.

'That's good', I pointed out.

'Yes', Besti *Behn* said. 'Just imagine if such a thing were to happen before the groups were formed! Either the child would have died because we had no money. Or we had to go to the moneylender and he would have taken away our entire year's crop for giving the money. That would have meant migration for at least one year'.

'Tell me something', I asked her. 'Is it that because of what you do in the group, you have more control over your lives?'

Bhuri *Behn*'s face brightened. 'Yes. We, Bhil women, led a wretched life. We worked much harder than the men. We did all the domestic chores. And we had to work in the fields too. But when it came to taking decisions, nobody asked us anything'.

'Have things changed now?' I asked her.

She nodded. 'Yes. I now have money in the bank. I now know what to do with it. And I know that I'll use it for good things. I know what is good for me. And what is bad. I can speak my mind now. If you had come here before and asked me questions, I wouldn't have talked like I am doing now. I would have kept quiet. Somebody would have answered for me. Just see now. I have not stopped talking'.

'What about other Bhil women?' I asked.

She sat up straight. 'It's not only me. All the women in my group are like that now. We talk a lot. We are not afraid of saying things. If we think a bad thing is happening, we talk openly about it'.

'What do you think, Ratan?' I asked.

He touched his moustache. 'That's true. Earlier they wouldn't talk. All you could get from them was a nervous titter. But now that they have started talking, they won't stop'.

The villagers stated that two funds, namely Gramkosh and Vikas Khata, have been established in the two villages. The money from the Gramkosh is used to finance various developmental activities and also for refinancing for the funds of the Self-Help Groups. Funds from the Vikas Khata will be used for the maintenance of the watershed works after the project is completed. Villagers contribute 5 per cent of their earnings from the watershed works to the Vikas Khata.

According to the villagers, as a part of the watershed development project, awareness camps had been organised in the two villages on issues such as education, health and hygiene; the villagers had greatly benefited from these camps. As a result, there was a high level of awareness amongst the villagers on socio-political issues. The villagers had mounted pressure to ensure that the governmental functionaries like the health workers and teachers were regular in their visits to the villages.

However, they were not very confident of managing things once PRAYAS, the PIA, withdraws at the conclusion of the project.

Sanda village

In Sanda Village, Parvati *Behn* who headed one of the Self-Help Groups in the village—the Shriram Group was present during the interaction. There were other Bhil women and Mohan Mavji who was responsible for forming 20 Self-Help Groups in the two villages of Sanda and Dunglawani.

According to the villagers, organisations such as watershed association, Watershed Development Committee, user groups and Self-Help Groups had been formed and were functioning in Sanda village. All the households in Sanda are members of the watershed association. Almost all the households in the village had been given exposure to successful watershed projects, and were also given skill training. Most of the households in Sanda Village had actively participated in the planning and implementation of the watershed activities. Most households in the village have been organised into Self-Help Groups. There are 14 Self-Help Groups in Sanda village.

These groups have financed a large number of activities of the members. They have given loans to the members for seeds, fertilisers and other agricultural operations. They have also financed income-generating schemes such as poultry, rope making, and even for purchasing a thresher for hiring out to others and generating income. Poultry, in particular, seemed to be very popular in Sanda village; being an established tradition and an important livelihood source in the Bhil community, this kind of entrepreneurship has high potential.

'What does your group do?' I asked Parvati *Behn*.
She chewed on her lower lip. 'We give loans. For seeds. For fertilisers. For schemes which make money. I've taken a loan for poultry'.
'Are you going to make some money?' I asked her.
'Of course, I will. You see, I've purchased a hundred chickens. I hope to make a profit of at least Rs 35 from each chick. Do you know what I did? I went and talked to a woman in Badi Malpur who had taken a loan from her group and raised chicks. She had made a lot of money. She has paid back her loan and is now going to raise another 100 chicks. She only told me what kind of chicks to rear'.
'What kinds?' I asked her.
'Oh! All kinds. Even the *Kadaknath* variety'.
'Do you eat meat?' I asked her.
She made a face. 'No, we don't touch it! We are vegetarians'.

Dependence of the villagers on moneylenders had come down because the Self-Help Groups are in a position to meet the credit needs of the villagers. As the watershed project has created jobs, migration has also come down. The institutions created by this project function in a manner so as to empower women and give them a role in the decision-making process.

'Now tell me something, Parvati *Behn*', I asked. 'You say you're a vegetarian. And yet, you're raising chickens. Where do you do it?'

'In a part of our house only', she said.

'Doesn't your husband object?'

Her face darkened. 'Why should he? He knows we'll make money out of it. And the money will be as much his as mine'.

'Did you ask him, though?'

She rubbed her temple and frowned. 'Yes, I told him. Well, I know why you're asking me that question. Yes, he took all the decisions in the past'.

'What kind of decisions?'

'Like going to the moneylender to borrow money. And when to migrate and where'.

'You go to the moneylender now, do you?'

'Not any more', she said and there was a hint of pride in her voice. 'Why should we? We can take loans from the group, can't we?'

'Yes, of course', I said. 'Let me ask you something else. What about migration?'

'We don't migrate now. Not after the works have started in our village. But, before that, we used to migrate every year'.

'Is that so?' I asked.

Parvati *Behn* wrinkled her brow. 'Yes, every year. Ever since I was a child. That time, I went with my parents. When I got married, I migrated with my children. This migration, it's hard on a woman. I can tell you that'.

'Is there a lot of work now in the village?'

She smiled. 'Yes. The watershed works. And it is all our work. We only decide what work to take up'.

Both Parvati *Behn* and Mohan Mavji felt that it will be difficult to replace PRAYAS as far as the development of the village was concerned.

Mankankvi village

At the time of the visit, Gajri, Jelu and Hakri, the women presidents of the three Self-Help Groups of Makankvi or Mankankvi village, were

present. According to them, institutions such as watershed association, Watershed Development Committee, user groups and
Self-Help Groups had been formed for the village. All the households
of Mankankvi village had been organised into these institutions, and
they had actively participated in the formulation and implementation
of the watershed action plans.

Self-Help Groups had been very active in mobilising savings of
their members and using the money to meet their credit needs. The
groups had also mobilised finance from external sources such as the
DWACRA scheme of the government and the banks and refinance
facilities from NABARD.

Self-Help Groups had even taken the initiative of starting *falia*
schools. As a result, the rate of education had gone up in the
village. This was so particularly for the girl child. The school drop-
out rates had also decreased. Several awareness camps had been
organised in Mankankvi village, and because of these camps, the
awareness levels of the villagers had increased, particularly in respect
of awareness of the importance of vaccination and family planning.
The villagers were now in a position to demand better and more
regular services from the government functionaries. As a result,
the teachers were more regular, the health officials came to the village
more promptly, and on the whole, the quality of public service had
improved.

Gajri said. 'Most Bhils are illiterate. We Bhils are very scared of the
written word. That's why we avoid places where reading or writing
is needed'.

'Are you still afraid of the written word?'

Gajri held up her hand. 'No, not any more. Not after the groups have
started working. We are still illiterate, but we are not afraid now. We ask
somebody to explain it to us. We now understand many more things
that we didn't understand before. Education is very important. We want
our children to go to school. Particularly our girls'.

'What's to be done for that?' I asked her.

She chewed on her lower lip. 'We have to open more and more
schools. In each *falia*, there should be a school. So that girls don't have
to walk big distances to go to school. Do you know something? Our
group has already started a *falia* school'.

'That's very good!' I pointed out.

'Oh, we now understand the importance of education', Gajri said. 'In
our village, we see to it that every child goes to school'.

According to the villagers, the Watershed Development Committee had taken up a large number of watershed works in the village. The works included treatment of private and common land, gabion structures, check dams, ponds and wells. Most of the villagers had been given training in watershed and exposure to other watersheds. This project had also given them a sense of awareness about ecological degradation, and the steps to be taken to overcome it.

They were of the view that the works taken up by the Watershed Development Committee were qualitatively better than the works taken up by the Panchayat. They were also participatory in nature. The panchayat took unilateral decisions about the public works in the village, which were not in the best interests of the villagers and also, shoddy in quality.

According to the villagers, things have improved a great deal in Mankankvi village after the watershed project was implemented. The water table in the village had gone up and the villagers were able to take on a *rabi* crop. Because of the employment provided by the project, migration has come down and there has been less dependence on the moneylender. The establishment of the grain bank by the villagers under the project has ensured that the Bhil in the village did not starve in the lean months as they used to do before.

'Have things improved for the village after these watershed works?'

'Very much so', Gajri continued. 'There is so much water now. You should see the water in our wells. It has gone up. You can almost put your hand down and touch the water. It's so high!'

'What about the crop?'

Gajri smiled. 'It's much better now. Do you know that we can even take on a *rabi* crop. And also vegetable crops. But the best thing is that there is so much more work in our village now'.

'Has the migration stopped?'

'Not totally', Gajri replied. 'Some people still go. But it has reduced considerably. You can count the number of migrants on fingertips. Now, tell me, why should we go to far away places when there is so much work here?"

'That's true', I agreed. 'Do the Bhil have enough to eat throughout the year?'

Gajri sighed. 'Yes, now we have. But we had problems in the past. What we got from our fields didn't last us for the whole year. We had to go to the moneylender and borrow grains. And what a lot of interest the moneylender charged! And we had to sell our next crop

(Box Continued)

> to him very cheap. No, that kind of thing doesn't happen anymore'.
> Hakri spoke up. 'Why don't you tell him about the grain bank the group
> runs, Gajri?'
> Gajri nodded. 'Oh, yes. Our group runs a grain bank. The Bhil can now
> borrow grains from the grain bank. They don't go to the moneylender.'
> 'Tell me something, Gajri', I asked. 'This project—watershed, Self-Help
> Groups, watershed committees—has it given you a lot of power?'
> She gave me a thoughtful look and then, smiled. 'Well, let me see.
> Yes, the project has done a lot of things for us. We don't starve during
> the lean phase. We don't have to migrate to far away places in search
> of work. There is work here. Our crops are better. We don't have to
> go to the moneylender. We save money now. We can take a loan
> from the group and start something. You see, I've taken a loan for a
> smokeless *chullah* (oven) and a small grocery shop.'

The villagers were of the view that the watershed project had made a great difference to their lives. There was less violence in the village, and the Bhil in the village had given up drinking. The villagers of Mankankvi did not go out at night and commit highway robberies like they used to do before. It has also led to empowerment of the Bhil women.

The villagers were uncertain as to what will happen if NCHSE, the PIA for Mankankvi village for the watershed project, withdrew from the activities at the end of the project. There was a distinct feeling that there would be setbacks.

Badi Dhami village

At the time of the visit, those present included Paskeli, Bhuri *Behn*, Anna, Tara, Kali and Soroj—all presidents of Self-Help Groups of Badi Dhami village, and Shri Petik and Laloo of the Watershed Development Committee.

> 'How many Self-Help Groups do you have in this village?' I asked
> Anna.
> 'There are six groups of women', she said. 'And two groups of men'.
> 'How many members in each group?' I asked her.
> 'About 10 to 20 members'.
> 'What do you do in the groups?'
> 'We save money. Each member saves from Rs 10 to Rs 20 every month'.

'What do you do with the money you save?'

It was Paskeli who answered my question. 'We give it out as loans to our members. What we save is not enough for that. So we get money from the bank. We show our savings to the bank and the bank gives us money. We also get loans from the Gramkosh. And we have got some grant from the Collector. Under the DWACRA scheme'.

'Loans for what?' I asked.

Paskeli smiled at me. 'Oh, for all kinds of things. Like seeds, fertilisers. Even for dug wells'.

'What's the recovery like?'

'Very good. Nobody defaults. Only in some cases, the members ask for more time to pay the loan back. The group listens to the defaulting member and takes a decision'.

'Why don't you go to the bank and take a loan directly? Why from the group?'

Anna spoke up now. 'The group is so much better. It's our own group, isn't it? If you go to a bank, there are so many problems. Lots of paper work. Lots of questions. And palms to be greased'.

According to the villagers, institutions such as watershed association, watershed development committee, user groups and Self-Help Groups had been formed for Badi Dhami village. Almost all the households, including women and the poor of the village, had been organised into these institutions and were active participants. The villagers had been given training in watershed and had been taken on exposure visits to successful watersheds in the district and even outside. According to the villagers, the Self-Help Groups had been particularly active in village development in Badi Dhami.

According to the villagers, the Watershed Development Committee had taken up many watershed activities which included a dam, a big plantation of bamboo and cultivation of grass in the common land of the village. They had been actively involved in the formulation of the action plan and its implementation. They were also aware of what caused ecological degradation and what steps were to be taken to prevent it.

'Tell me, what did you do to make the watershed programme work?' Tara gestured to Bhuri *Behn*. 'Let Bhuri *Behn* tell you that. After all, she is the one who went to Ralegaon Sidhi. Didn't you, Bhuri *Behn?*' Bhuri *Behn* said with an embarrassed smile, 'I don't know why everyone picks on me because I went to Ralegaon Sidhi! All of them know what we have to do. All right, all right. Let me explain'.

(Box Continued)

> 'Well, we have to level our land', she said in a high-pitched voice. 'Cultivate only across the slopes. We have to do bunding. What is that bunding called?'
>
> 'Contour bunding', I suggested.
>
> Bhuri *Behn* looked gratefully at me. 'Yes, that's what it is. We have to plant trees on those bunds. We have to collect the monsoon rain and store it. We have to control the grazing of our cattle and do it on rotation basis. And we have to grow grass or fodder. Isn't that it?'
>
> Bhuri *Behn*, in her own simple way, had described the scope of watershed project more succinctly than I would ever do.

The villagers said that they had benefited from the implementation of the Watershed Development Project. They have no shortage of fodder, thanks to the grass grown on the common land. There is more water in the village and this has enabled them to take a second crop.

> 'What do you do in the common land?' I asked.
>
> 'In the common land, we grow bamboo', Paskeli said. 'We have also planted grass there. To grow fodder for our cattle'.
>
> 'Do you have enough fodder now?'
>
> It was Kali who replied. 'More than we need. We even sell fodder'.
>
> 'How do you manage that?' I asked her.
>
> 'Let me tell you. All the people in the village cut the grass together and collect it in a place in the village. The Watershed Development Committee calculates the requirement of fodder for all the families in the village and the grass is distributed according to that'.
>
> 'But you said you also sold a part of it', I said. 'isn't it?'
>
> 'Yes, that's true', Kali agreed. 'We have decided that half of the grass that we get from the common land should be sold and the money deposited in the Gramkosh. We do that'.
>
> 'That's fantastic', I said. 'Now, has your committee taken up watershed works in the village?'
>
> 'Quite a few', Kali said.
>
> 'Have you people benefited from these watershed works?'
>
> 'A lot', Kali assured me. 'Let me tell you what has happened. The water level in the village has increased. You should see our wells. They are full now. And you should see the common land that we have treated. Old roots of the trees that the Bhils had cut 50 years back are coming back to life'.
>
> 'That's very impressive', I suggested. 'Now, what about your crop? Is there an improvement?'
>
> 'Yes. We can even take on a *rabi* crop now. And we have managed to get our own lands treated also'.

According to the villagers, there is much more employment available in Badi Dhami village itself because of the watershed works. As a result migration has come down. The grain bank that has been started by the villagers as a part of the Watershed Project has ensured that the villagers do not go to the moneylender to borrow grains. The Self-Help Groups have been in a position to mobilise savings of their members and extend credit both for meeting credit needs and consumption requirements.

According to the villagers, this project has led to empowerment of the Bhil in Badi Dhami. The villagers when asked what will happen when CBCD, the NGO that is the PIA for the Watershed Project in Badi Dhami village withdrew at the conclusion of this project, were confident that they would be able to manage.

9

Performance of the Project

The last chapter was about people's perception: of those who planned and executed the project. From the people's response, it looked as if the project had succeeded in what it set out to do. The water table had risen, trees had been planted, supply of fodder had increased, and the Bhil villages grew more self-sufficient. But how did the project fare in physical terms? In this chapter, we look at the findings of a study that carried out an intermediate assessment of the impact of the watershed project (Rajiv Gandhi Watershed Mission 1999), a satellite evaluation of the impact of watershed project in selected locations, the performance of the watershed villages during the drought of 1999–2000, and the final project outcomes in 2005.

Intermediate assessment (1999)

Wasteland area

The study looked at the position of wasteland in 11 micro watersheds for the years 1994–95 and 1997–98, and found that the area of wasteland had decreased in each of these micro watersheds (see Fig 9.1).

The area of wasteland decreased from 12 per cent in 1994 to 4 per cent in 1998 (see Fig. 9.2). This represents a decrease of 66 per cent and the total area reclaimed is of the order of 798 hectares in 11 micro watersheds.

The decrease, according to the study, was due to a combination of factors: social and vegetative fencing and physical measures such as cattle protection trenches and stone bunds. Soil and water conservation measures such as gully plugs and staggered contour trenches had increased soil accumulation, reduced the rate of run-off, and increased the recharge of groundwater. Reseeding had established vegetative cover on bunds and trenches while providing for better quality fodder.

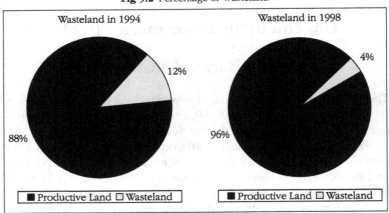

Fig. 9.1 Wasteland Area

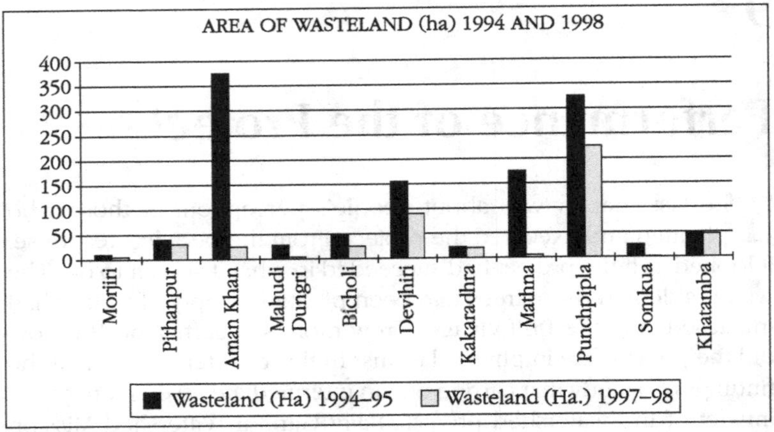

Source: Rajiv Gandhi Watershed Mission (1999)

Fig 9.2 Percentage of Wasteland

Source: Rajiv Gandhi Watershed Mission (1999)

Reduction in soil erosion/silt accumulation

This study found that a targeted approach to achieve effective, efficient and site-specific soil and water conservation had been adopted. Ridge to valley treatment had been strictly followed.

Contour trenches, staggered pits, gully control measures, drainage line treatment along with fodder development and plantations of suitable species were taken up on the recharge zone. Contour bunds, earthen embankments, *nalla* bunds, sub-surface dykes and other water conservation and harvesting structures had been taken up in transition and discharge zone. Agronomic measures like inter-cropping and intensive cropping along with pasture development and *silvi-pastoral* activities had been taken up in transition and discharge zone. Intensive treatment had been completed on 13,430.50 hectares.

The survey looked at the treatment work completed, and the findings are set out in the following chart (see Fig. 9.3).

The chart shows the depth of silt deposition measured in June 1998. The figures are in centimetres and an average of depth taken from 10 sites per micro watershed. Greater rates of silt deposition are observed in areas with less vegetative cover. While high figures represent a reduction in soil loss, they also show a greater proportion of denuded land.

Fodder development

The survey found that fodder development had been promoted in private as well as government land. Grassbeds, *silvi-pasture* and pasture development had been given emphasis. The study sampled 2,950 hectares of land which had been taken up for fodder

Fig 9.3 Silt Deposition : Mean of 10 Sites per Village

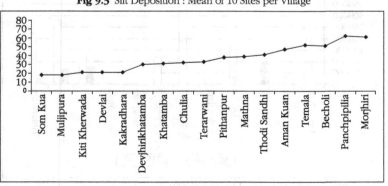

Source: Rajiv Gandhi Watershed Mission (1999)

development. Dinanath, Sukli and Batodi were the major fodder crops taken up. It assessed the availability of fodder in 12 micro watersheds. Most of the watershed villages had become self-sufficient in fodder and forage. Fodder development had provided immediate returns to the villagers, and also, the environ for biomass generation and soil conservation.

This chart (see Fig. 9.4) shows the change in position of availability of fodder in 12 micro watersheds. The improvement, according to this study, was as a result of protection of forest and wasteland, reseeding with improved grass species, and adoption of improved grazing management regimes. It also found that increased harvest of grass represented a significant gain in economic terms, as chronic fodder shortages prior to protection meant that many households were forced to buy fodder from the market, often far away from the village.

Plantation, seed sowing and regeneration of degraded forests

This study found plantation in private and community lands had been given priority among watershed activities. To supplement these activities, soil working and seed sowing in vast expanses had been

Fig 9.4 Fodder Development

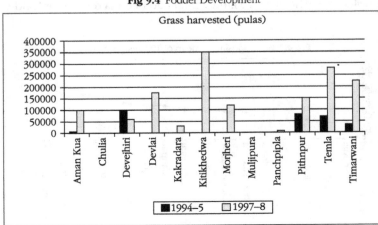

Source: Rajiv Gandhi Watershed Mission (1999)

taken up. The saplings for plantation activities had been raised in kitchen gardens, in nurseries developed by members of Self-Help Groups, in Self-Help Group nurseries, and in nurseries raised by PIAs. To supplement plantation activities, soil working and seed sowing had been taken up in vast areas.

The study looked at regeneration activities in 14 micro watersheds, and found that activities for generation from dormant seeds, coppices and gap planting had been taken up in degraded forest areas (Fig. 9.5). The village forest committees under the Joint Forest Management Scheme had provided the necessary protection. According to this study, plants had regenerated in 19,354 hectares of degraded forest in 1998 in these 14 micro watersheds as against 4,933 hectares in 1994. The study noted that the type of protection was a combination of cattle proof trenches, stonewalls, vegetative fencing and social fencing. Approximately 32 lakh plants were found to be regenerating in watershed areas through social fencing by the Watershed Development Committees. Tectona Gradis (teak) and Butea Manosperma (palash) were the major copicing species. This study found the following regeneration by land type (Table 9.1).

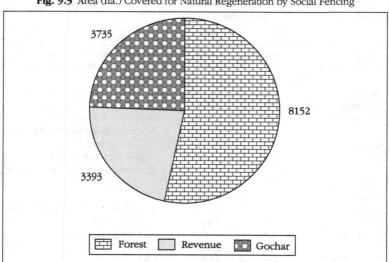

Fig. 9.5 Area (ha.) Covered for Natural Regeneration by Social Fencing

3735

8152

3393

☐☐ Forest ☐ Revenue ☒ Gochar

Source: Rajiv Gandhi Watershed Mission (1999)

Increase in water resources

This study (see Fig. 9.6) plotted the change in water level for the period between 1994 and 1998. This was done by looking at the level recorded in the observation wells.

The increase in water level is the difference between the level in June 1994 and June 1998. The increase, according to this study, was due to the enhanced recharge in the upper catchment and prolonged flow of surface water as a result of soil and water conservation measures. The study also gathered information about people's perception of the period of flow, and plotted the change, as recorded in the chart(see Fig. 9.7).

The study measured the change in the availability of water in 20 micro watersheds. The chart (see Fig. 9.8) shows the total number of water resources.

This study found that the Bhil women did not have to trudge long distances for fetching water; this was due to the fact that there was an increase in the number of wells, and water was available nearby. According to this study, 63 wells which used to dry up before the end

Table 9.1 Regeneration of Plants

Micro Watershed	Per Ha. No. of Plants Regenerated		Type of Protection
	1994	1998	
Chulia	605	605	No protection on common land
Kiti-Kherwada	300	1,0 0 0	CPT, Stone wall social fencing, MFP, etc.
Morjiri	7	78	
Pithanpur	80	1,700	
Muljipura	1245	1,333	Vegetative fencing, CPT, Stonewall social fencing, MFP, etc
Kakadhra	500	4,000	CPT, Stone wall
Aman Kuan	165	315	Vegetative fencing, CPT, Stonewall social fencing, MFP, etc
Devlai	100	400	
Bicholi	10	20	
Devjhiri	0	3,070	Social fencing, Vegetative fencing
Mathna	0	623	Social Fencing
Puchpipla	450	760	CPT, Vegetative fencing
Temla	712	2,800	CPT, CCT
Thod Sindhi	759	2,650	CPT, CCT

Source: Rajiv Gandhi Watershed Mission (1999)

Fig. 9.6 Increase in Water Table in Blocks of Jhabua District between 1996 and 1997

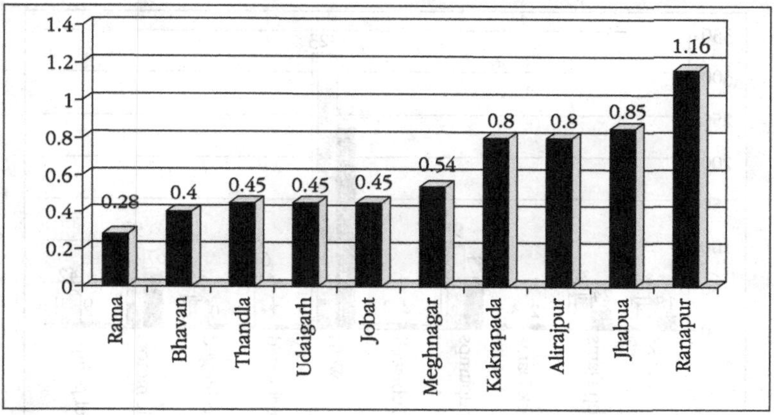

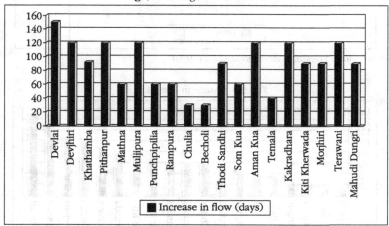

Source: Rajiv Gandhi Watershed Mission (1999)

Fig. 9.7 Change in Water Level

Source: Rajiv Gandhi Watershed Mission (1999)

of summer had now water throughout the year because of increased recharge of groundwater. Also the change in water level in 19 micro watersheds as recorded in the observation wells were measured. The following chart gives the details. It is clear that there had been a change in water level, although the increase varied across and between the micro watersheds (see Fig. 9.9).

Fig. 9.8 Water Resources 1994 and 1998

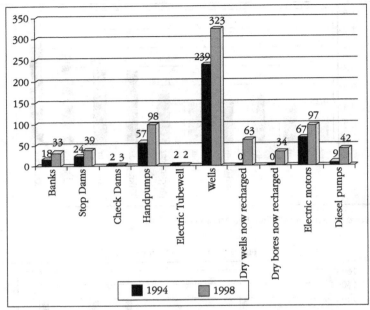

Source: Rajiv Gandhi Watershed Mission (1999)

Fig. 9.9 Change in Water Level

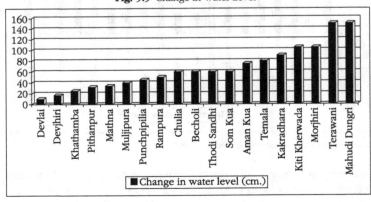

Source: Rajiv Gandhi Watershed Mission (1999)

According to the study, 15 percolation tanks had recharged the groundwater and they had also provided water for the livestock.

There was an increase in sub-surface moisture because of more vegetation near the tanks. The study found that because there was more water, there were a number of lift irrigation schemes, either installed or renovated, and also an increase in the number of water-lifting devices such as electric motors and diesel operated pumps. The study collected information regarding irrigated area from revenue records and interviews for years 1994–95 and 1997–98. In all watersheds, the area had increased over the period of the mission. This was directly attributed to light irrigation schemes, but the study found that the increased water availability was a result of water conservation measures. The increased area under irrigation had led to increased cropping intensity changes in cropping pattern and increased security in case of drought.

Improved cropping practices

Area of crops for both *kharif* and *rabi* seasons for the years 1994 and 1998 was collected only in respect of seven micro watersheds, because the Bhil did not keep records; neither did they use standard measurements. The area was calculated using the seed rate as a proxy (see Fig. 9.10).

As the data show, there was an increase in the area of cash crops such as soyabean and cotton; in fact, between the two, their share

Fig.9.10 *Kharif* Crops

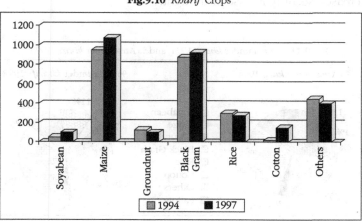

Source: Rajiv Gandhi Watershed Mission (1999)

of the cropped area had gone up from 3 per cent to 8 per cent. The area under maize, the staple crop of the district, had also increased: from 29 per cent of the cropped area, it has gone up to 31 per cent. The following charts depict the change, providing a comparative picture of the crops under *kharif* for 1994 and 1998 (see Fig. 9.11).

The charts provide an interesting insight: the share of the fallow land decreased from 19 per cent to 12 per cent. The reasons for such a decrease, according to the study, are mainly two.

(a) Increased availability of human resources i.e., labour and (b) Increased access to low-cost credit as a result of the *Bairani Kuldis* funding the credit requirements of the Bhil.

The cropped area under *rabi* increased from 1,286 hectares to 1,626 hectares; an increase of 340 hectares in seven micro watersheds. As the charts show (see Fig. 9.12), there was an increase in the area of wheat—from 8 per cent to 12 per cent—which, as the study found, was due to the increased availability of surface water and access to water-lifting devices.

On the whole, the cropped area had increased in all the seven watersheds. The increase in cropped area was because of the fact that more water was available, and this, in turn, was the result of water conservation measures. There was also an increase in the area under irrigation. The increased area under irrigation had led to increased cropping intensity, changes in the cropping pattern, and increased security in case of drought.

Fig 9.11 % Area under *kharif* 1994 and % Area under *kharif* 1998

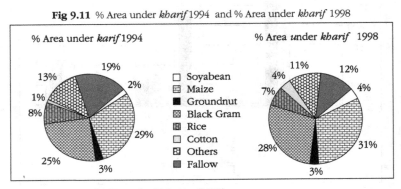

Source: Rajiv Gandhi Watershed Mission (1999)

Fig. 9.12 Area under *rabi* Crop

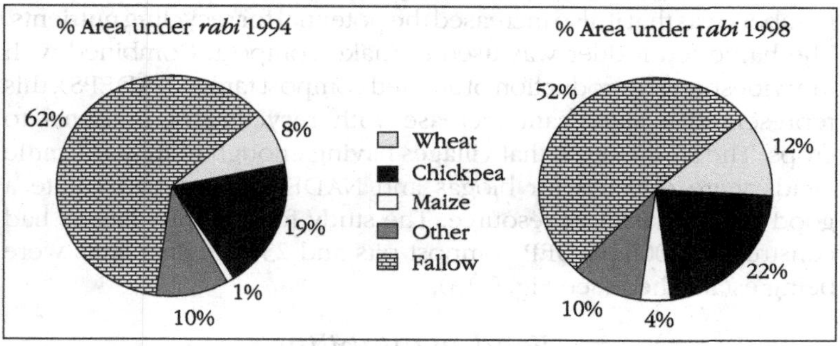

Source: Rajiv Gandhi Watershed Mission (1999)

Yield of crops

The study found that there was an increase in the yield of major crops on an aggregated average over 1995–96 and 1996–97. The trend showed a general increase in production of important crops. There was a significant increase in the yield of maize, wheat and soyabean; wheat and soyabean, in fact, registered phenomenal increases over the previous years. There was some decline in the yield of crops like peas, blackgram, *moong* and groundnut. The study attributed the decline to pest resurgence and untimely rains. For example, the sharp decrease in the yield of black gram was caused by late rains in the *kharif* season and high levels of insect infestation. On the whole, there was an overall increase in crop production per unit; the increase, according to the study, was due to increased access to cheap and timely credit resulting in increased fertiliser use.

Large ruminants

This study found that in response to increased availability of fodder, the number of large ruminants had increased. The increase upscaled livestock numbers to the optimum for productive and reproductive purposes. Some villages where fodder became abundant saw an increase in the number of buffalo for milking purposes and there was already an evidence of incorporating primary milk cooperatives in these watershed villages. This study found no evidence to suggest an increase in the off-take of small ruminants, which were kept for different reasons. An important aspect of the increase in livestock numbers was that it also increased the potential for recycling nutrients. The harvested fodder was used to make compost. Combined with the widespread introduction of aerated compost tanks (NADEPS), this represented a significant increase with recycling of nutrients to crops. The study found that villages having enough number of cattle heads were selected for biogas and NADEP in order to create a good organic fertiliser resource. The study found that villagers had constructed 2006 NADEP compost pits and 231 biogas plants were being established (see Fig. 9.13).

Food availability

This study assessed the position in respect of availability of food in 16 villages and found that it showed an increasing trend. It also computed availability of food in terms of the number of months for which foodgrains lasted for an average Bhil household (see Table 9.2).

Clearly, there was an increase in availability of food. This increase, according to this study, was due to the increase in the cropped area and in the yield of the crops.

Migration

It collected data on migration in respect of 10 micro watersheds, and as the data in the table (see Table 9.3) show, there was a reduction in the number of families migrating and the period of migration.

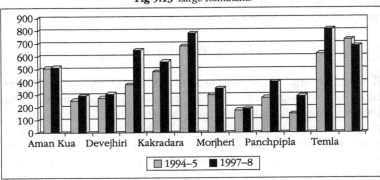

Fig 9.13 Large Ruminants

Source: Rajiv Gandhi Watershed Mission (1999)

Table 9.2 Food Availability by Watershed

Villages	Food Availability (months)	
	1994	1998
Chulia	4	5
Kiti-Kherwada	8	12
Morjhiri	6	8.5
Temarwani	5	7
Muljipura	6.5	7
Aman Kua	10	11
Mahudi Dungri	5	9
Devlai	6.5	11
Bicholi	5.5	7
Devjhiri	7	11
Kakradhara	6	12
Mathna	5	6
Panchpipaliya	6	7
Themla	9	11.5
Thod Sindhi	7	10
Somkua	8	10

Source: Rajiv Gandhi Watershed Mission (1999)

Table 9.3 Migration of Households

Villages	Migration				
	Household			Months	
	1994	1998		1994	1998
Chulia	60	33		5	3.5
Kiti-Kherwada	30	20		8	4
Morjhiri	83	67		8	6
Pithanpura	77	87		6	6
Temarwani	110	78		5	4
Aman Kua	89	98		9	6
Mahudi Dungri	60	10		6	4
Devlai	37	17		6	5
Devjhiri	50	21		6	3
Temla	70	30		4	3

Source: Rajiv Gandhi Watershed Mission (1999)

The data establish that migration decreased in all the villages except Pithanpura, where the number of migrating households went up marginally, while the number of months remained the same, and in Aman Kua, where the number of migrating households went up but the number of months came down. The reduction in the number of migrating families and the period of migration, according to the study, was due to increased opportunities for wage labour created locally by the watershed project. This project increased the quantum of work in the farming sector. In addition, increased availability of food led to greater self-sufficiency and as a result, there was less of a compulsion to migrate.

Dependence on moneylenders

It looked at the changes in the credit pattern in 11 villages and the findings are set out in the table (see Table 9.4). This study also shows that there was a decrease in both the number of households borrowing from moneylenders and the size of the loans.

The reduction in the number of loanees and the amount of loan, according to the study, is attributed to the fact that the compulsion to borrow had decreased as the income from various sources including the cash for wage labour had increased. It was particularly true of loans for food grains, which were required towards the end of summer when the availability of food grains is rather low.

The study found that there were Self-Help Groups in almost all the villages, providing credit for agricultural inputs and loans for any other purpose that the group members approved of. These groups,

Table 9.4 Changes in Credit

Watershed	H/h Borrowing		% Reduction in Loan Amount
	1994	1998	
Chulia	67	48	N.A
Morjhiri	113	110	10
Pithanpura	112	87	20
Temarwani	146	123	N.A
Aman Kua	319	213	35
Devlai	86	43	15
Devjhiri	48	10	35
Kakradhara	85	54	N.A
Panchpipaliya	100	75	25
Temla	36	36	N.A
Somkua	35	20	15

Note: N.A – not available.
Source: Rajiv Gandhi Watershed Mission (1999)

this study noted, set their own lending rates which were between 2 and 8 per cent per month whereas the moneylenders charged up to 12.5 per cent per month. It also noted that freedom from the clutches of the moneylenders was considered an important thing in the Bhil community as the interest rate charged by the moneylender was exorbitant.

Bairani Kuldi

The study looked at the functioning of the *Bairani Kuldis*. In the watershed areas across the district, the process of formation of Bairani Kuldi had been uneven; the number was as low as 53 groups in Ranapur block while it was as high as 292 in Meghnagar block, as the chart below (see Fig. 9.14) indicates.

There were 2,181 *Bairani Kuldis* in the district. They had 30,681 members, the average membership of a group being 15. Although the bulk of the funds in the *Bairani Kuldis* had come from institutional sources, as the chart (see Fig. 9.15) shows, the level of savings of the members of the groups had been as high as 25 per cent of the total funds at the disposal of the *Kuldis.*

The details regarding the membership and the number of loans given by the *Bairani Kuldis* are set out in the chart (see Fig. 9.16). The credit given by the *Bairani Kuldis* included loans for agricultural inputs like fertiliser, seed and irrigation, as well as consumption credit.

Fig. 9.14 Number of *Bairani Kuldi* per Block

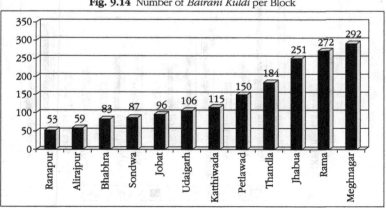

Source: Rajiv Gandhi Watershed Mission (1999)

According to this study, the informal credit rates of the *Bairani Kuldis* had entered production costs; in addition, they had also ensured food security in the domestic portfolio.

Fig. 9.15 Funds in *Bairani Kuldis*

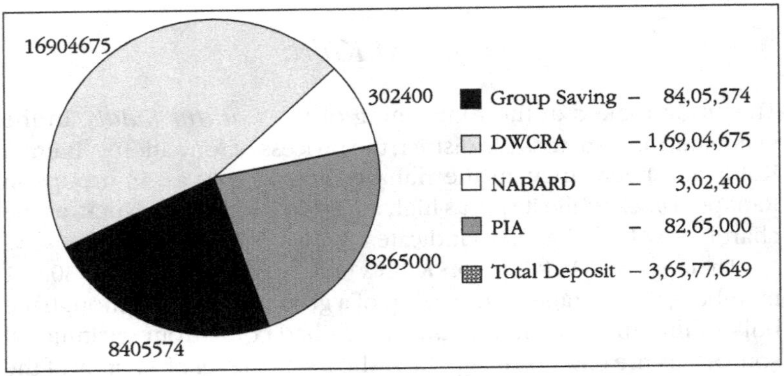

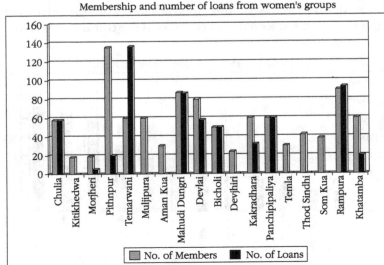

■ Group Saving	–	84,05,574
☐ DWCRA	–	1,69,04,675
☐ NABARD	–	3,02,400
▨ PIA	–	82,65,000
▦ Total Deposit	–	3,65,77,649

Source: Intermediate Assessment of Watershed Management (1999)

Fig. 9.16 Membership and Number of Loans

Membership and number of loans from women's groups

Source: Intermediate Assessment of Watershed Management (1999)

It also found that the Bhil women, while managing the *Bairani Kuldis*, had been exposed to the processes of keeping accounts, charging interest rates, asking for linkages to increase their capital stock, and pressurising for recoveries.

Non-farm activities

There was a significant increase in non-farm activities. These activities included nursery raising, sericulture, dairy, poultry, fisheries and a number of income generating activities supported and funded by the Self-Help Groups. The increase in non-farm activities meant that the income in a Bhil household was no longer dependent on agriculture alone; it was now also from off-farm activities and as a supplement to the agricultural income. This was a welcome development in Jhabua district where droughts are a recurring feature and single-point dependence on income from agriculture could be disastrous.

Training and exposure

The study found that the process of training and exposure in the project had given the members of the Bhil community adequate awareness about watershed, and also about their roles and responsibilities in the project. According to it, the training requirements were diverse, but sincere efforts had been made in the project to develop modules to meet those needs.

The mission had organised various training programmes: Workshop on remote sensing by MPCOST; Participatory Research Appraisal; Preparation of action plan; Accounts related training modules for WDC's; watershed development concept; plantation and fodder development; WDC secretaries; workshop on agricultural input and credit; Workshop on fish farming and duckary; women saving banks; training through satellite on medicinal plants and their uses; and camps and training on cattle development and management.

Change in net income of a village

The study also evaluated the changes in the net income that had accrued as a result of the implementation of the watershed project. It also looked at the changes in the net income experienced in Themla village which is located in the southern part of the district.

The data in the table (see Table 9.4) shows that the net income from all sources had increased for the medium farmers—they experienced an increase in overall net income of the order of 35 per cent. The bulk of this increase came from agricultural income and from the common property resource, as the chart (see Fig. 9.17) makes it clear.

According to this study, the increase in income from common property resources was related to increased collection of fodder and non-timber forest products. But the increase in income from agriculture was due to more complex reasons which are listed below.

- The increase in availability of fodder meant that a larger livestock population could be sustained, and in most cases, there was an increase in the number of large ruminants—buffalo in particular.

Table 9.5 Change in Net Income

Change in Net Income	Medium		Better Off	
	+/– Rs	+/– Rs	+/– Rs	+/– Rs
Agricultural	3,337	26.8%	5934	26.7%
CPR	1,100	82.5%	–95	–4.9%
Local Labour	486	182.0%	306	71.7%
Migration	308	43.4%	–783	–86.0%
Overall	5,231	35.4%	5362	21.0%

Source: Rajiv Gandhi Watershed Mission (1999)

Fig. 9.17 Change in Net Income by Source

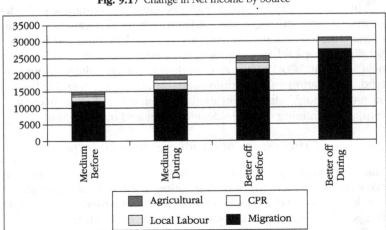

Source: Rajiv Gandhi Watershed Mission (1999)

Table 9.6 Accrual of Benefits

Percentage of Earnings	Medium		Better Off	
	Inc Before	Inc During	Inc Before	Inc During
Agricultural	84.3%	79.0%	87.1%	91.2%
CPR	9.0%	12.2%	7.7%	6.0%
Local Labour	1.8%	3.8%	1.7%	2.4%
Migration	4.8%	5.1%	3.6%	0.4%

Source: Rajiv Gandhi Watershed Mission (1999)

More large ruminants translated into higher milk production and occasional sales of offspring.

- The increase in the number of large ruminants enhanced the capacity to recycle nutrients in the farming system i.e., by consuming crop by-products, their dung enriching the compost, and returning more of the nitrogen, phosphate and potassium which might otherwise have been lost. Enhanced fertility of soil led to increased crop yields.
- New seeds of high yielding varieties had been introduced and this has meant higher productivity.
- New lift irrigation schemes had been started in the village which had further enhanced agricultural production.

One of its significant finding is that such benefits accrued to both the better-off and medium households by almost equal percentages, as the table (See Table 9.5) shows.

This study found that farmers with larger landholding had benefited by a greater margin in cash terms. But, for both the categories—the medium households and the better-off—the bulk of the income had come from the local natural resource base, and in Themla, migration formed only a negligible part of the household income. Such a finding may not hold good in the case of other villages in Jhabua district where the extent of common land is much less, and the density of population is greater, leading to a heavier reliance on migration.

Satellite evaluation of the impact of watershed project

The Remote Sensing Applications Centre, Bhopal carried out an evaluation study of the impact of the watershed project in Jhabua district in August 1998 (Impact Evaluation Using Remote Sensing Techniques 1998). For this purpose, nine milli watersheds, one each in Bhabra, Jhabua, Rama, Meghnagar, Thandla, Ranapur, Jobat, Alirajpur and Sondwa blocks were taken up for computer-based data analysis. The milli watersheds were selected on a random basis. The objective of this study was to assess the pre- and post-implementation status of the selected milli watershed using remote sensing technique. The idea was to monitor the impact of the watershed project in terms of land use and land cover parameters in respect of the cropped area, wasteland, water-bodies and biomass.

Data from the Indian Remote Sensing Satellites—1B, 1C and 1D—were used for the study. Satellite imagery for the period December 1993 and January 1994 was used as the base-year data for the pre-implementation period, and the same period data of 1997 and 1998 were used to depict the status for post-implementation of the project. Analysis of block-wise monthly rainfall data showed that the years 1993–1994 and 1997–1998 had normal rainfall; a fact which confirmed that the impact of rainfall in the watersheds during the selected years could be considered as normal.

The technique used for the analysis of data was the conventional technique used by the Indian Space Research Organisation for such evaluation studies. The milli watershed boundaries were digitised and the satellite imagery was put to analytical rigour; the milli watershed boundaries were ported on the satellite images in order to get milli watershed-wise extracted images of 1993 and 1994 pre-implementation phase and 1997 and 1998 post-implementation phase for purposes of comparison. Visual comparison of these images gave an estimate of the overall changes that had taken place in these selected milli watersheds, while, at the same time, locating these changes on the image with precision.

The images were classified in order to identify various land use and land cover categories, and calculate the geographical area for comparison and detect changes that have taken place. All land use and land cover classes such as standing crop, water bodies, wasteland (land with or without scrub), fallow land (agricultural land without standing crop), forest-type vegetation and vegetation cover (tree cover in the milli watershed irrespective of forest land) were identified and classified. Visual comparison of these classified images gave the locations where changes in land use and land cover categories had taken place. Comparison of area statistics gave the changes in terms of increase or decrease in the geographical area of various land-use and land-cover categories. Normalised Difference Vegetation Index (NDVI) analysis was done to study the vegetation dynamics. NDVI images are used for discriminating vegetation vigour; they are generated showing five classes of vegetation vigour: very good, good, medium, low and very low. Visual comparison of NDVI images depicts the location of areas where the crop or forest vigour has improved. Comparison of satellite images, NDVI images, classified images, and area statistics indicated the overall changes in land use and land cover categories, status and changes in vegetation vigour, and changes in geographical area of land-use and land-cover categories.

Observations in Bhabra block

Visual comparison of 1994 and 1998 images showed that there was a considerable increase in the extent of redness, indicating increase in crop-land area and vegetation vigour. The water spread, as well as the number of water bodies, had increased over a period of four years. The classified images showed that the extent of crop land, mainly the double-cropped area, had increased by 1,112.19 hectares. The increase in the extent of crop land was seen around the villages of Mathana, Bada Bhavta, Biljhar and Chhoti Kareti. The water bodies in the south of the villages of Khakari, Kuva and Mathana were not observed in 1994, but these were clearly seen in 1998 images. The statistics showed an increase in the area of waterbodies by 56.23 hectares. The fallow land, seen near village Sejawada and to the north of the village Kanasanungaliya in 1994, had been converted into good crop land as observed in 1998 image. The extent of this land had been reduced by about 166.50 hectares in the entire milli watershed. A part of the wasteland had been reclaimed and its area had been reduced by 874.58 hectares. The NDVI images depicted an overall increase in the vegetation vigour and biomass, especially around Bhabra and Roligam settlements, indicating considerable increase in productivity.

Observations in Jhabua block

Visual comparison of 1994 and 1998 satellite images showed an increase in the extent of redness, indicating an increase in the area of crop land. These changes were clearly visible around Kheri, Kundla and Mahuri villages. An evaluation of December 1993 and December 1997 images in Hathipawa Reserve Forest showed an increase in the extent of smooth, textured, brownish red coloured patches, which indicated an enhancement in fodder resources in terms of grass and bushes in the area. These changes were not very clear in January/February images due to drying out of grassy/bushy vegetation during this season. Wasteland and fallow land had been reclaimed in several places. In the extreme northern part of the milli watershed, the forest-type vegetation had increased around village Kankaradara, due to conservation and regeneration activities. A new water body with considerable water spread was observed in 1997 and 1998 images in the west of Kankaradara village, which was not seen in 1993 and 1994 images. Area statistics obtained from classified images of 1993 and 1998 showed an increase in the crop land by 759.07 hectares and an increase in the waterspread by 130.90 hectares. The NDVI images showed an overall increase in vegetation vigour especially around Pitol Chhoti, Naldi, Dumpara villages and also in the western part of the milli watershed. The overall changes at various places in the milli watershed were progressive and productive.

Observations in Rama block

Visual assessment of December 1993 and December 1997 images showed that forest-type vegetation in Rama and Umarkot located in the northern part of the milli watershed had improved, which was depicted by a reddish colour in 1997 image. These changes were not very clear in images of January and February, 1998, perhaps due to leaf fall during the period. An increase in the extent of redness along drainage was observed near Ruparel, Hatyadeli, Chhapri and Machhlia villages. This indicated an increase in the extent and improvement in the condition of the crop land. Patches of fallow land that existed between Bhamariya Pipaliya and Machhlia villages and also in the north of village Machhlia in 1994 were converted to crop land in 1998. The wasteland, at places, had been reclaimed and converted into fallow land, as observed especially in the southern part of the milli watershed located towards north of the village Margarundi. The land use and land cover statistics derived from the supervised classification of 1994 and 1998 images showed an increase in the crop land by 188.58 hectares, increase in the spread of waterbodies by 63.85 hectares, and increase in the area of fallow land by 254.71 hectares. The wasteland had been reclaimed and reduced by 701.55 hectares over the four year time period.

The NDVI images indicated an increase in the good and very good vegetation vigour classes and transformation of low biomass cover to medium biomass cover over this period. The temporal change analysis of the watershed suggested an overall improvement in terms of vegetation cover and reclamation of wasteland. Satellite evaluation showed that, there was improvement in terms of increase in forest-type vegetation and vegetation vigour, increase in the spread and number of waterbodies, and enhancement of fodder resources in terms of grass and bushes. Wasteland had been reclaimed and reduced, and there was an increase in extent of crop and fallow land. The assessment showed that the overall changes brought about by the watershed project were progressive and productive.

Performance during the drought of 1999–2000

A severe drought visited Jhabua in 1999–2000. This was one of the worst droughts in the history of the district: the rainfall was just about 500 mm as against the normal rainfall of 800 mm, and that too, intermittent with long, dry spells; 1,254 villages out of 1,374 in Jhabua district registered less than 37 paise *anavari* (it is a system by which the loss of crop is measured as the fraction of a rupee). The production of *kharif* crop dropped from an annual average of 2,59,000 tonnes to a mere 98,000 tonnes. The area sown under the *rabi* crop plummeted from 1,25,000 hectares to 37,000 hectares. There was 60 per cent reduction in the cropping area, and production declined by 25 per cent. But the destruction of the *kharif* crop meant loss of the maize crop—the staple diet of the Bhil. About 25 per cent of the villages faced acute drinking water shortage because of the drought. All in all, it was a bad drought (Collector, Jhabua 2005).

How did the watershed project help?

The watershed project helped in several ways at the time of the drought of 1999–2000. The water conserved on the surface ensured relatively better crop—both during *kharif* and *rabi*, and the sub-surface water ensured more moisture, helping the *rabi* crop. There was enough water for the cattle. There was water in the wells and handpumps for drinking purposes. The fodder development activities ensured adequate fodder for the cattle in the watershed villages. The grain banks delivered food grains to the needy Bhil. The *Bairani Kuldis* ensured thrift and credit to their members, and livelihood activities such as dairy, fisheries, horticulture and *gramodyog* supported sagging incomes. Use of hybrid seeds especially in the maize crop meant higher production. Regeneration of teak and allied forest ensured better fuelwood supply in the watershed villages. Drip irrigation sets distributed in the watershed programmes helped backyard horticulture; 277 units irrigated 120 hectares of horticultural crops.

Watershed villages versus normal villages

One gets a clearer picture when a comparison is made between the conditions prevailing in the watershed villages and the non-watershed villages at the time of the drought of 1999–2000. The *kharif anavari* in the watershed villages was of the order of 34 to 37 paise, while in the neighbouring non-watershed villages, the *anavari* for the *kharif* crop was 4 to 5 paise less. The *rabi* area in the watershed villages was 15,000 hectares while for the district as a whole, the *rabi* area was only 37,000 hectares. (The normal *rabi* area of the district is 1,20,000 hectares while that of the watershed villages is 37,000 hectares). In the watershed villages, fodder production was 76.92 lakh bales during 1999–2000, while the non-watershed villages had to import fodder. In respect of drinking water, 2,389 out of 2,639 handpumps were working in the watershed villages (90 per cent) while in the non-watershed villages, only 6,918 out of 8,494 handpumps worked (80 per cent). Cost of digging wells in the watershed villages was less, as water was available at lower depths.

Final project outcomes (2005)

Natural resource base

There are significant gains in the natural resource base in the watershed area, as the chart (see Fig. 9.18) indicates.

Fig. 9.18 Water Resources

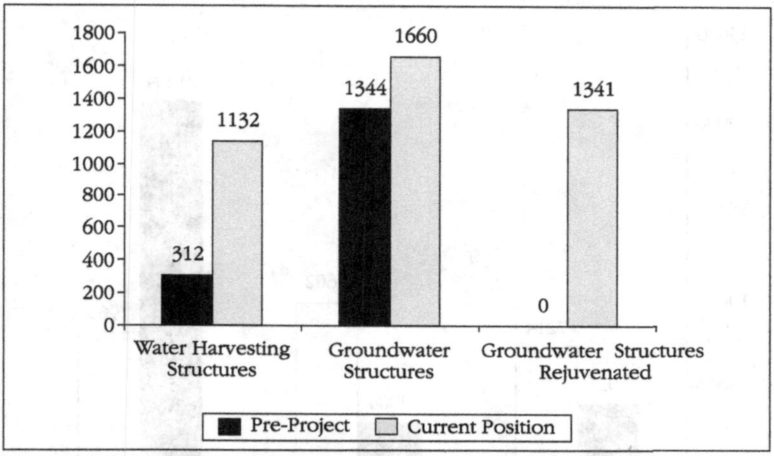

Source: Collector Jhabua (2005)

Clearly, there is an improvement in the water resource base. This means that much of the rainfall and the available run-off in the watershed is conserved. The increase in the water resource base has brought welcome relief to the Bhil: it has meant availability of drinking water for human beings and cattle. The drudgery faced by the Bhil women in fetching drinking water over long distances is reduced.

Irrigation, forest and wasteland

The area under irrigation and forest has increased, and the area under wasteland has decreased, as shown in the chart (see Fig. 9.19).

The area under irrigation has almost doubled. The area under forest cover has increased more than three times, and there has been a significant reduction in the wasteland area. On the whole, there is an increase in biomass production. There is a qualitative enhancement of the natural resource base: the natural resources of the watershed such as land, water and vegetation have been optimally utilised. This has mitigated the possible adverse effects of drought, prevented further ecological degradation and contributed to restoring the ecological balance in the area.

Fig. 9.19 Irrigation, Forest and Wasteland

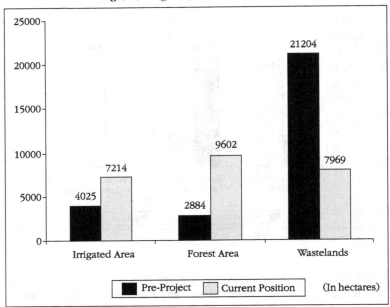

Source: Collector Jhabua (2005)

Agricultural production

The chart (see Fig. 9.20) provides details of agricultural production in the pre-project period and the current position.

There is an increase in both *kharif* and *rabi* area. Particularly impressive is the addition to the *rabi* area. The increase in the area of *kharif* and *rabi* crops means two things.

- The vulnerable *kharif* crop that requires survival irrigation in a long dry spell now gets that life-saving irrigation.
- The *rabi* crop that needs residual moisture and assured water is in a position to get them.

There is also an increase in the yield of crops for both *kharif* and *rabi*.

Fig. 9.20 Agricultural Production

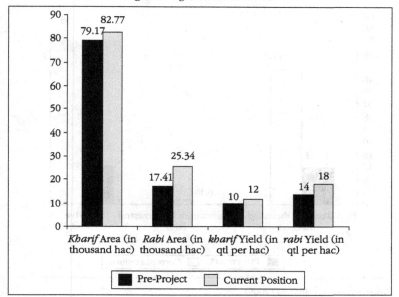

Source: Collector, Jhabua (2005)

Plantation activity

The details of the plantation activities in the watershed area are set out in the chart. (see Fig. 9.21).

On the whole, there is an increase in the area both under fuel wood and horticulture. This has three important implications:

- A large part of the common property resources (revenue, forest and panchayat land) is covered under vegetation through plantation and natural regeneration.
- The households are able to meet their fuel requirement from local sources.
- The drudgery faced by Bhil women for fetching fuel wood is reduced.

The chart (see Fig. 9.22) notes the improvement in the availability of fodder.

Fig. 9.21 Fuel Wood and Plantation

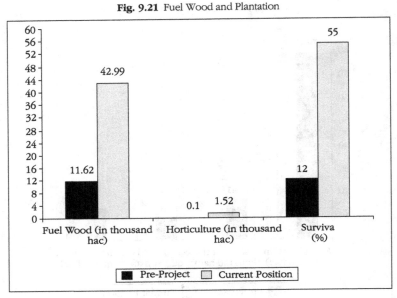

Source: Collector, Jhabua (2005)

Fig. 9.22 Fodder Availability

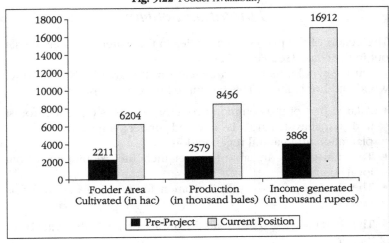

Source: Collector, Jhabua (2005)

The area under fodder cultivation has improved significantly. The boost in the availability of fodder means that the families in the watershed area are able to meet their fodder requirement from local sources. It also meets a long-standing need of the Bhil villages; in fact, one of the expectations in the project was that most of the watershed villages would have become self-sufficient in fodder and forage by the time of the completion of this project.

Another expectation in this project was that fodder development would provide immediate returns to the villagers. This has happened. The income generated from fodder has gone up, but, more importantly, the availability of fodder from local sources means that the hapless Bhil women do not have to undergo the drudgery of fetching fodder over enormous distances.

The increase in the availability of fodder means that a larger livestock population is sustained, and therefore, there is an increase in the number of large ruminants. There is also a marked increase in stall feeding from the pre-project days: it has increased from 10 per cent in the pre-project period to 35 per cent at present (Collector, Jhabua 2005). In fact, it was the common experience before the project started, that regeneration of grazing land was much less than consumption. Since there was a shortage of fodder, animals were not stall-fed and were let out to graze; this meant that there was no control over cattle grazing. According to experts, if there is more stall-feeding and the grassland is protected for a period of three to five years, the grass would regenerate, because the root stock still remains firmly in the sub-soil and regeneration is possible if the conditions are compatible.

Non-farm activities

Non-farm activities have also been taken up in a big way in the watershed area (see Table 9.7).

The increase in non-farm activities means that the income in the Bhil household is no longer dependent on agriculture alone: it is also from off-farm activities and as a supplement to the agricultural income. This is a welcome development in Jhabua district where droughts are a recurring feature and single-point dependence on income from agriculture has proved to be disastrous.

Table 9.7 Non-Farm Activities

Activities	Achievement
Nursery Raising	105 No.
Sericulture	74 hac
Dairy	504 milch cattle inducted
Poultry	42,000 bird broilers and Kadaknath provided to SHGs
Fisheries	164 structures covered
Income-generating	39 activities started by
Activities	465 Bairani Kuldi groups

Source: Collector, Jhabua (2005)

Community organisations

The performance of the project in establishing community organisations is set out (see Table 9.8).

The commendable thing about the project is the way it has created, nurtured and sustained institutions at the village level that are managed by the Bhil themselves. These community organisations, by their operation often linked with income-generating activities, have benefited a large number of Bhil.

Convergence with economic and social development programmes

There is a convergence of economic and social development programmes in the watershed area. This is evident from the implementation of the different schemes of the government as the table (see Table 9.9) shows.

Table 9.8 Community Organisations

Groups	Constituted	Benefited	Linked with Income Generation
User Groups	2276	20010	45
Self-Help Groups	1423	14213	192
Women Thrift& Credit Groups	1205	14984	232

Source: Collector, Jhabua (2005)

Table 9.9 Different Schemes of Government

Sector	Scheme	Achievements
Agriculture	Nadep	2,478 No
	Biogas	249 No
	Lift Irrigation	5 No.
	Bio Fertilisers	762 Kg
	Inter Cropping	Increase in per hac yield Induction of Soyabean, Cotton, Wheat, Groundnut.
Horticulture	Vegetable Production Developed	Kitchen Gardens
	Kalptaru	250 hectares
	Horticulture Garden	1,424 hectares
	Drip Irrigation	82 No
	Ber Budding	1.82 lakh
Education	Padna Badna	All SHGs are covered
	Falia Schools	62 No
	Regular Attendance & New Admission Campaigns	Door to door contact, Rallies, Books, Dresses & School Bags provided through Gramkosh
Health	Immunisation	100%
	Family Planning	279 Nos
	Cataract	192 Nos operated
	Leprosy	53 Nos treated
	Tuberculosis	109 Nos treated
	Health Camps	59 Nos
	Cattle Camps	82 Nos
	Anaemia Control	1,786 No treated
	Malnutrition	105 No treated
	Shouchalaya	810 No constructed

Source: Collector, Jhabua (2005)

The reason for such convergence are the awareness camps organised to educate the Bhil on the importance of education, health, hygiene, agricultural productivity, etc. as a part of the watershed project. As a result of these awareness campaigns, the awareness level of the Bhil has increased, and they are in a position to demand and get the benefits of the developmental schemes of the government.

10

Assessing the Performance

In order to undertake a critical assessment of the performance of the project, it is necessary to refer to the Logical Framework Analysis that had laid down, at the time of the formulation of the project, the intended outcomes or the results to be achieved. It had also stipulated some success indicators to assess the quality of the outcomes. The projected outcomes were divided into four categories: environmental, economic, social and institutional.

Environmental

In respect of the environment, the outcome had been defined as the optimum utilisation of the watershed's natural resources such as land, water and vegetation which would mitigate the adverse effects of drought and prevent further ecological degradation leading towards restoration of ecological balance. The Logical Framework Analysis had stipulated the following success indicators for assessing the achievements in respect of environmental outcome:

a. Increase in groundwater table;
b. A good part of the *kharif* cultivation receiving life saving irrigation in a long dry spell;
c. *Rabi* area requiring residual moisture (gram) to be under cultivation and parts of the *rabi* area to be able to produce an irrigated crop in a drought year; and
d. All families in the watershed able to meet fuel and fodder requirement from local sources even under drought condition.

Increase in the ground water table

The observation wells which record the level of water on a regular basis showed considerable increase in the water level. The increase can be attributed to the enhanced recharge in the upper catchment and prolonged flow of surface water as a result of soil and water conservation measures and construction of stop dams. People's perception of the period of flow in streams which was gathered in group discussions, also showed that there was increase in water resources.

Kharif cultivation receiving life-saving irrigation

Wells are the primary source of water in Jhabua district. As we have seen, there was increased water in the wells because of increased recharge from groundwater. In fact, as the intermediate assessment study pointed out, after four years of the watershed project, a total of 63 wells, which had previously dried up before the end of summer, had water now throughout the year because of increased recharge from groundwater. The increased water in the wells was used for the survival irrigation of crops during long, dry spell in the *kharif* season. We have also seen how during the severe drought of 1999–2000, the increased level of water resources saw through the *kharif* crop during the dry spell.

Rabi crop during a drought year

While discussing the ravages caused by the drought of 1999–2000, we have seen how the water conserved on the surface ensured a good crop, both during *kharif* and *rabi*, and that the sub-surface water ensured moisture, thus helping the *rabi* crop. We also found in the course of our interactions with the Bhil in the villages that they were able to take a fully irrigated *rabi* crop after the watershed project was implemented.

Meeting fuel and fodder requirements

The intermediate assessment study showed that there was a change in the availability of fodder as a result of the protection of forest and wastelands and reseeding with grazing management regimes. The study also showed that the watershed villages had become self-sufficient in fodder and forage. During interactions with villagers, it was indicated that the Bhil were able to meet their fuel and fodder reqirements locally, and were, in fact, in a position to sell the surplus grass, with substantial income flowing to the Gramkosh in the process. Even during the bad drought year of 1999–2000, the fodder development activities undertaken in the watershed villages ensured adequate fodder for the cattle locally.

Economic

In respect of economic activities, the outcome had been defined as increased farm (agriculture and allied) productivity leading to improved and sustained livelihood status of the watershed community with special emphasis on poor and women. Here are the stipulated success indicators for assessing the achievements in economic outcomes:

a. Increase in the household income;
b. Employment generated locally as a result of watershed development and augmentation of natural resources;
c. Increase in irrigated area;
d. Decrease in dependency on moneylenders for productive and consumption loans; and
e. Household income generated through on-farm and off-farm activities as supplementary to agricultural income.

Increase in household's income

As was indicated in the Themla study and also during interactions with the Bhil, there has been an increase in the income of the households. The increase is primarily from agriculture: there is increase in the double-cropping area, the *rabi* irrigation potential and the survival irrigation for *kharif*; the cropping intensity has increased; and there has been a change in cropping pattern from food to commercial. There was also an increased income from common property resources by way increased collection of fodder and non-timber forest products. There is also an increase in household income from on-farm (poultry, dairy, goatry) and off-farm (brick/basket/leaf plate making) activities as supplementary to agricultural income.

Employment generated locally

Interactions with the Bhil indicated that adequate employment was being generated locally. Data on migration generated by the intermediate assessement study corroborates this. There has been significant reduction in migration in the watershed area, both in respect of the number of families migrating and the period of migration. This reduction is directly attributable to the opportunities for wage labour created locally by the watershed project. Other activities of the watershed project such as higher labour inputs for increased crop production, harvesting of fodder and collection of non-timber forest produce from the common property resources and the forest, have also increased opportunities for local wage labour. On the whole, there has been a significant increase in employment generated locally as a result of watershed development and augmentation of natural resources.

Increase in irrigated area

As the intermediate assessment study shows, there has been a substantial increase in the area under irrigation. There is increase in both *kharif* and *rabi* area as well as an increase in the yield of crops. Appropriate land management practices are being followed in the watershed area: intercropping and line sowing, and more importantly, contour cultivation. The practices adopted are as per the gradient and other attributes of the land.

Dependency on moneylenders

As the data of the intermediate assessment study show, there has been a reduction in the number of households who borrow from moneylenders in most watershed villages. It is also significant that the size of the loans has decreased between 10 per cent and 35 per cent. This is attributable to the income that has accrued to the Bhil households from the increased receipts from wage labour and agricultural productivity. As was pointed out by the Bhil, there has been a significant reduction in loans for food grains from moneylenders because of the activities of the communally managed grain banks. In addition, because of the operations of the *Bairani Kuldis*, which have provided loans for productive consumption and income-generation purposes, the dependency on the moneylenders has been greatly reduced. This is particularly important in the Bhil setting because the moneylenders charge up to 12.5 per cent interest per month while the lending rates of *Bairani Kuldis* have been between 2 per cent and 8 per cent per month.

Income from on-farm and off-farm activities

Interactions with the Bhil indicated that the watershed project had created opportunities for income from both on-farm and off-farm activities. The increase in off-farm activities has meant that the income of a Bhil household in the watershed area is no longer dependent on agricultural activities. This is particularly commendable in the Jhabua setting where sole dependence on agriculture has been damaging in the past during times of drought.

Social

In terms of the social aspect, the outcome was defined in the project as the creation of an empowered community with an increase in the level of awareness on social issues, and women and poor being given due share in the development process of the society. The Logical Framework Analysis had stipulated the following success indicators for assessing the social outcomes: rate of education increased; school drop-out rates minimised; awareness level increased for the importance of vaccination and family planning; awareness increased about health, education and other services and the quality of services ensured; reduced level of liquor consumption; decreased number of conflicts; and drudgery faced by women for fetching drinking water, fuel wood and fodder reduced.

Rate of education increased

Interactions with the Bhil indicated that the rate of education among the Bhil children, and particularly among the girls, had increased. They, in the watershed area, are now very conscious of the fact that

Bhil girls should attend schools and get educated. This is an important development in the context of the traditional perception of the Bhil that there was no point in sending a girl to school because she was going to get married. On the whole, the increase in the rate of education amongst the Bhil has been helped by the fact that the local community has been goaded to action to open schools in each *falia*, making it easier for Bhil boys and girls to commute to school.

School drop-out rate minimised

The immediate assessment study came to the conclusion that there has been a significant decline in the school drop-out rate in the watershed area. This is also borne out by interactions with the Bhils. This was helped by the fact that there were campaigns for regular attendance and new admission campaigns. There were door-to-door contact and rallies as well as distribution of books, dresses and school bags provided through Gramkosh.

Awareness level increased for the importance of vaccination and family planning

There is an increase in awareness level in the project area regarding health and hygeine. This is evident from the fact that there is cent per cent immunisation in the watershed area; family planning camps have been a success; and cases of cataract, leprosy, tuberculosis, anaemia and malnutrition have been treated; health camps have been conducted; and *shouchalayas* have been constructed. That there has been increased awareness for the importance of vaccination and family planning was evident from interactions with the Bhil.

Awareness increased about health, education and other services, and the quality of services ensured

Regular awareness camps have been organised in the watershed area to educate the Bhil on the importance of education, health and hygiene as a part of the watershed project. As a result of these aware-ness campaigns, the Bhil now demand regular visits of health workers, teachers and other departmental staff in their villages. Because of the spread of awareness, the Bhil are aware of the importance of getting the benefits of the developmental schemes of the government and ensuring the quality of the services delivered.

Reduced level of liquor consumption

Interactions with the Bhil indicated that the level of liquor consump-tion in the watershed area has gone down significantly. In village after village, it was the common refrain in the interactions that the Bhil had vowed to give up drinking altogether and social sanctions were in place against consumption of liquor. This was as a result of sustained campaign against liquor organised at regular intervals.

Decreased number of conflicts

Interactions with Bhil also indicated that there was a general de-crease in the number of conflicts. It was also suggested that the commission of crimes such as murder, highway robbery and acts of vandalism had come down drastically. This was, in part, attributable to the reduced level of liquor consumption.

Drudgery faced by women for fetching drinking water, fuel wood and fodder reduced

The intermediate assessment study indicated that water level in the wells had increased. This has spared the Bhil women the drudgery of fetching drinking water from long distances. Fodder development taken up in private as well as common property resources has increased fodder availability and all the watershed villages are now self-sufficient in fodder and forage. The protection of forest and wastelands and adoption of improved grazing management regimes have meant increased availability of fuel wood, and this, in turn, has reduced the drudgery of the Bhil women in fetching fuel wood.

On the whole, the drudgery-reducing interventions in the watershed project have helped the Bhil women to find the time and energy to take part in their strategic roles and responsibilities in watershed development, and thrift and credit activities.

Institutional

In respect of the institutional aspect, the outcome had been defined as sustained community action for the operation and maintenance of assets created and further development of the potential of the natural resources in the watershed. The Logical Framework Analysis had suggested the following success indicators for assessing the institutional outcomes.

a. Assets created under watershed to be maintained and benefits accrued in the some proportion.

b. Watershed Development Committees to mobilise additional resources from line departments for further development of natural resources.

c. Credit needs of the community are met through self-help groups and Gramkosh.

d. Adequate availability of financial resources through local mechanisms (Self-Help Groups or Gramkosh) to meet the credit needs of the households.

e. System to be in place for meeting agricultural needs to minimise the effect of middlemen.

f. Increasing trend in Watershed Development Committees, user groups and Self-Help Groups from observer–discussant–decision maker.

g. Women and poor have access and control over the benefits accrued through watershed activities.

Maintenance of assets and accrual of benefits

Institutionally, the responsibilities for maintenance of assets have been given to the Watershed Development Committee, and more instrumentally, to the user groups. Financially, this has been ensured by the constitution of the Development Fund at the level of each watershed. The Development Fund is the futuristic fund set up with contributions from the Bhil for maintenance of infrastructure created by the project. An amount of Rs 210 lakh is available in the Development Funds of 587 villages in the watershed area. The expenditure of money from the Development Fund can be author-ised by the respective Watershed Development Committees. This means that for the maintenance of the assets which have been created by the watershed project, the Bhil do not have to be at the mercy of the government; the maintenance can be done by utilising the money available in the Development Fund duly authorised by the Watershed Development Committee. Regarding the equitable sharing of the benefits accruing from the watershed assets, this was assured by giving the responsibility to the Watershed Development Committee, which was representative in character and functioned democratically.

Mobilising resources from the line departments

The project design ensured this. But, in the actual implementation of the project, the watershed institutions were articulate enough to demand and get the additional resources from the different schemes of the various departments of the government for further develop-ment of natural resources in the watershed. This was also done by utilising the funds provided in the different schemes of the Cen-tral Government and the Government of Madhya Pradesh for develop-ment of natural resources under the watershed project.

Credit needs of the watershed community being met through Self-Help Group sources and Gramkosh

The intermediate assessment study indicated how the *Bairani Kuldis* in the watershed area had met the credit needs of the house-holds by providing loans not only for agricultural inputs such as fertilisers, seeds and irrigation, but also underlining their distribu-tional implications by providing consumption credit. This was also borne out by interactions with the Bhil. The work done by the

Bairani Kuldis in fulfilling the credit needs of their members was an outstanding achievement, and particularly so in the totally degraded environment of Jhabua district, where indebtedness is chronic and almost every Bhil household is caught in an inexorable debt trap. The important thing is that the *Bairani Kuldis* have succeeded in freeing the Bhil from their lifelong covenant of debt with the moneylenders. The funds available in the Gramkosh were used for refinancing the loans given by the *Bairani Kuldis*.

Adequate availability of financial resources through local mechanisms

The *Bairani Kuldis* increased their capital stock by networking with government schemes and formal financial institutions so that the credit needs of their members could be fully funded. The magnitude of the financial resources that the *Bairani Kuldis* mobilised is impressive: It is of the order of Rs 3.66 crore–Rs 3 lakh from NABARD, Rs 82 lakh from the PIAs, Rs 169 lakh from the DWACRA scheme, and Rs 112 lakh of savings of their own members. (Collector, Jhabua 2005). The record of recovery of the *Bairani Kuldis* has been equally impressive—it has been consistently more than 90 per cent. Working with a single-minded determination and putting the group dynamics in place, the Bhil women in change of the *Bairani Kuldis* recovered the loans back with a professionalism that should be the envy of any formal financial institution. These Bhil women were not heartless either; in deserving cases they allowed time for the repayment of loans, without losing sight of the goal that the money of the group needs to be recovered. The Gramkosh consisting of contributions made by the villagers from payments received by watershed wages provided refinancing facilities to the Self-Help Groups.

System in place for meeting needs of agricultural production to minimise the effect of middlemen

The watershed project put in place a system for meeting the needs of agricultural production without going through intermediaries. For example, the Self-Help Groups came together to form federations which provided them linkages with government programmes, bulk supply and purchase, provision of services, policy advocacy, etc. The examples are the Laxmi Mahasamiti and the Mahila Mahasamiti as we found out in our interactions with the NGOs. These federations were in a position to buy fertilisers and other agricultural prerequisites at a price cheaper than the market and supply them to the farmers, thereby minimising dependence on middlemen.

Increasing trend for the Bhil from observer–discussant–decision-maker in the institutional deliberations

Interactions with the Bhil indicated that to start with, most of the Bhil were in the nature of observers in the deliberations of the community organisations. But as a result of exposures they got by way of training and awareness campaigns, the Bhil became discussants and ultimately, decision-makers. As it came across very clearly from interactions with the Bhil, they are now full participants in the decision-making process in the community-based institutions. With their participatory mode of functioning, the institutions at the village level now have the capacity of identifying and addressing the development needs of the Bhil community. The way these community organisations have functioned has been democratic, as all the members have participated fully in the deliberations on an equal footing; no patronage relationship has characterised the functioning of these community organisations.

Women and poor have access and control over the benefits accrued through watershed activities

As interactions with the villagers revealed, the Bhil women have been full participants in the deliberations of the watershed institutions, and have, in fact, taken a leading role in matters relating to village development. In this respect, the experience of Bhil women in forming and managing the *Bairani Kuldis* has been remarkable. The uneducated and ignorant Bhil women, while managing the *Bairani Kuldis*, have been exposed to the complex process of keeping accounts, charging interest rates, asking for linkages to increase the capital stock, and pressurising for recoveries. The educative yet demanding process of managing the *Bairani Kuldis* have thrown up new leaders, new expertise and new initiatives. For the Bhil women, the process has lowered social barriers that flow from distinctions of gender by enhancing their capacity. The association of Bhil women with the process of management of *Bairani Kuldis* has given them clout in the Bhil community, their own savings have given them economic independence, and their access to the resources of the *Bairani Kuldis* has given them power. The watershed project has brought the women to the vanguard of the Bhil community and has given them voice. The same is true of the poor in the Bhil community. The watershed project has been instrumental in lowering the social barriers that result from distinctions of social and economic status. The women and poor of the Bhil community, working through watershed institutions, have gained access to and control over the benefits accruing through watershed activities.

On the whole, the balance sheet of the performance of the watershed project in respect of the outcomes turns out to be positive. In respect of all that the vision of the project had propounded—environmental, economic, social and institutional—the implementation of the watershed project has delivered outcomes that fully meet the expectations of the project. There are two aspects, however, which were not contemplated in the project design, but flow out of the implementation of the project, that have the potential to impinge on project outcomes. These two aspects are the relationship with the civil society intermediaries and Panchayati Raj institutions.

Relationship with civil society intermediaries

A critical component of the watershed project was the association of civil society intermediaries (NGOs) in the conceptualisation, formulation and implementation of the project. The civil society intermediaries, as the Programme Implementing Agencies (PIAs) in the milli watershed, were given the responsibility for sensitising the Bhils and organising them into groups and community organisations.

They were expected to instil in the diffident Bhils a sense of self-worth, self-confidence and self-reliance. They had to help the groups in scaling up membership functions, awareness raising and training. They were also required to provide technical assistance and necessary skills to make the groups and community organisations viable and sustainable. They were also called upon to create bridges between the Bhil community organisations and the departments of the government and outside agencies. On the whole, the project assigned strategic roles to the civil society intermediaries in empowering the Bhil during the project period and enabling them to sustain the process after the implementation of the project was over.

In fairness, the civil society intermediaries performed their tasks exceedingly well. But in the process of doing so, they created a dependency which the Bhil find it difficult to shed. Ideally, there should have been an exit protocol written explicitly into the project design. As the interactions with the Bhil indicated, they will have a great deal of difficulty to get out of such dependency.

Relationship with Panchayati Raj institutions

The Panchayati Raj institutions are the officially designated agencies for village development. To that extent, they were given an important role in the institutions of the watershed project and represented in the Watershed Development Committees. It was also stipulated that the Gram Sabha of the panchayat should approve the watershed works to be taken up by the Watershed Development Committee and the Watershed Association. Interactions with the Bhils, however, indicated that they had no confidence in the Panchayati Raj institutions representing their collective interest. The panchayats and Gram Sabhas were portrayed as bastions of entrenched interests, which promoted the private interests of the Sarpanch at the expense of the good of the Bhil community. The perception of the ordinary Bhil was that the Panchayati Raj institutions did not advocate his interest, but because of elite capture, worked to his detriment. Interactions indicated an adverserial relationship between the watershed institutions and the Panchayati Raj institutions, and in the ultimate analysis, this does not portend well for the sustainability of either the watershed institutions or the gains that the watershed project has brought to the Bhil.

Empowerment

The important question is: has the watershed project in Jhabua empowered the Bhil? To answer that question, we have to assess the outcomes of the project in the context of the conceptual framework that we had drawn up in Chapter 4. In that chapter, we had defined what constitutes empowerment, indicated the institutional reforms necessary for empowerment and the elements of empowerment and drawn up a framework and identified major influencing conditions for empowerment in context. We had also stipulated what a watershed project should delineate by way of an empowering strategy.

Institutional reforms

In the conceptual framework, we had argued that a successful institutional strategy to empower the poor would depend critically on the institutional, social and cultural context. Institutionally, what is needed for the empowerment of the poor are institutions that are sensitive to their interests, and enable them to participate fully and meaningfully in the deliberations of these institutions and articulate their interests and aspirations. Socially and culturally, the empowering approach should put the poor at the centre of development and view them as the most important resource rather than as mere passive recipients, and build on local knowledge, skills, values, initiatives and motivation to solve problems and manage resources. We have seen how the Bhil organised themselves into community institutions, provided local leadership and participated in the planning and implementation of the watershed project. These community institutions enabled all participating Bhil, even the women and poor amongst them, to establish their role in village de-velopment. Working in these institutions, the Bhil identified the needs of their community and implemented measures to fulfil them.

Elements of empowerment

The conceptual framework had identified four elements, namely, access to information; inclusion and participation; accountability; and local organisational capacity. It had also stipulated that these four elements should act in synergy and not in isolation.

Access to information

In project formulation and implementation, the information flows to the Bhil have been impressive. The Bhil have been educated about the importance of watershed through rallies, meetings, Kala Jathas and street plays. Extensive land literacy campaigns were organised to give the Bhil an idea of the watershed resources, problems and possible solutions. Almost all the households in the watershed area were given exposure to successful watersheds. Awareness camps were organised on issues like education, health, hygiene, etc. at regular intervals. Armed with such information, the Bhil were in a position to take advantage of opportunities, gain access to services, exercise their rights, negotiate effectively and hold the institutions of state and society to account.

Inclusion and participation

As we have seen, all the Bhil in the watershed area were included in the membership of the community organisations for planning and implementing the watershed activities and have participated in the decision-making process such as agenda setting, defining priorities and allocating resources. Their full and meaningful participation in the deliberations of these institutions ensured that they had the authority and control over the resources at the village level, particularly the financial resources. This enhanced the capacity of the Bhil to engage in the Bhil society's power structure and allowed them to articulate their interests and aspirations.

Accountability

In the process of planning and implementating watershed works, the participating Bhil were in a position to hold the public function-aries accountable. In addition, because of the awareness created about health, education and other services of the government, the Bhil were enabled to ensure the regular visit of the health workers, teachers and other departmental staff of the governement. Because of the increase in their awareness level, they were in a position to demand and get the benefits of the various developmental schemes of the government.

Local organisational capacity

While planning and implementing the watershed project, the Bhil proved that they had the ability to work together, organise themselves and mobilise resources to solve problems of collective interest. Of course, they were helped in these tasks by civil society intermediaries who provided the crucial component of bridging and linking social capital. The civil society intermediaries sensitised the Bhil, mobilised them and put them in charge. The intermediation of the civil society intermediaries was crucial, but the fact remains that the Bhil were in a position to influence government decision-making and acquire collective bargaining power.

All these four elements promoting empowerment did work in synergy to make the community organisations of the Bhil the key players in the institutional context.

Major influencing conditions

Motivation

The Bhil in the watershed area were motivated to form institutions, which became the principal institutional mechanism for their empowerment. Working in these institutions, the Bhil understood the important role of these institutions and interactions in them, and the power that would accrue to them by asserting themselves as productive members of the society.

Self-management

The institutions that the Bhil formed and built elected their own leaders, evolved their own functional norms and developed working modalities. The institutions did succeed in managing their own activities such as conducting meetings, savings and credit operations, resolving conflicts and looking after common property resources of the community. The institutions were in a position to exercise full autonomy over the system.

Resource mobilisation

The institutions were also in a position to mobilise resources for funding their activities through regular savings of their members and raising capital stock from external sources. The pooling of such resources made the members realise how, when individual resources are combined, it enhances the individual position of each member. The groups also evolved a viable system of resource generation, allocation and utilisation in which benefits accrued to all members equitably, thereby ensuring ownership and sustainability.

Support

The institutions were in a position to secure support from civil society intermediaries in matters of technical assistance, skill building, scaling up their membership and range of functions, awareness raising and training. With such support, they were enabled to interact and negotiate with the organisations of state and society and outside agencies.

Decentralisation

By strengthening the capacity of the Bhil and building institutions at the local level, the project succeded in making it possible for the members of the Bhil community to participate in local governance. Such decentralisation made the departments of the government more responsive to the needs of the Bhil and was instrumental in achieving developmental goals in ways that met the needs of the local Bhil community.

Assets and capabilities

Individual

As far as material assets are concerned, we have seen how there was an increase: the incomes of the households increased; employment was generated locally as a result of watershed development and augmentation of natural resources; the area under irrigation increased; and significant household incomes were generated through on-farm and off-farm activities. Human capabilities such as the capacity for basic labour, skills and good health also increased. As a result of the community institutions built by them, the Bhil acquired social capabilities such as social belonging, relations of trust, sense of identity and the capacity to organise. They also acquired the capacity to represent themselves, form associations and participate in political life, because of the strength they derived from their participation in these institutions.

Collective

As we have seen, the project created and sustained community institutions in which the Bhil participated in the deliberations on an equal footing, and there was a progressive trend in the deliberations of these organisations: from discussants to decision-makers. Working through these community institutions, the Bhil, including the women and poor of the community, established their role in the decision-making process of village development.

On the whole, the watershed project expanded opportunities for the Bhil by creating assets and capabilities—both individual and collective—and also made it possible for them to increase the productivity of these assets and capabilities. In the process, the project gave the Bhil control over their lives and livelihoods by enabling them to take advantage of the opportunities for economic and social development.

Vulnerability

The Bhil community, as we have seen, is particularly vulnerable to two events—migration and recurring droughts. The watershed project adopted a comprehensive mix of measures to deal with them. As a short-term solution, the project helped the Bhil cope with immediate problems by way of creating opportunities for local employment and providing food security through the grain banks. In the long term, the project put in place an integrated livelihood strategy that succeeded in generating adequate household income through both on-farm and off-farm activities. The project also utilised the resources of land, water and vegetation optimally to mitigate the adverse effects of drought and prevent further ecological degradation, restoring in the process the ecological balance in the region.

The empowerment framework

In Chapter 4, we had diagrammed a conceptual framework that focused on institutional reform to invest in poor people's assets and liabilities, leading to improved development outcomes. We had indicated the following development outcomes:

a. Improved governance and acceess to justice;
b. Functioning and more inclusive basic services;
c. More equitable access to markets;
d. Strengthened civil society;
e. Strengthened poor people's organisations; and
f. Increased assets and freedom of choice.

In our analysis of the performance of the project, we had found that all these developmental outcomes had been achieved. There is improved governance, the basic services are now in a functioning mode and more inclusive, there is equitable access to markets, the civil society and poor people's organisations are strengthened, and there are increased assets and capabilities for the Bhil because of the watershed project, leading to freedom of choice for them.

Has the watershed project empowered the Bhil? We had defined empowerment as expansion of assets and capabilities of poor people to participate in, negotiate with, influence, control, and hold accountable institutions that affect their lives. Has the project done that? The answer is yes. The project has been instrumental in expanding the assets and capabilities of the Bhil—both individual and collective —in ways that has given them control over their lives and livelihoods, freedom of choice, and the ability to negotiate and influence relationships and decisions. The community institutions created by the project have enabled the participating Bhil—even the women and poor amongst them—to establish their role in decision-making and hold the institutions accountable. Thanks to the project, the Bhil now have a high level of awareness about social and political issues, a sense of economic self-confidence and a feel of self-worth. The project has given them new skills, new knowledge and new efficiencies. The Bhil have a voice in the institutions that shape their life. In that sense, the watershed project in Jhabua has created an empowered Bhil community.

Part-III

11

Conclusions

A nil Agarwal, a noted environmentalist is, alas, no longer with us. Though his life was so tragically brief, he fought fiercely to protect our environment. Before he died, he went around Jhabua looking at the implementation of the watershed project. He had this to say,

> What I saw was astounding. Instead of a moonscape, there was land being nursed and being brought back to life with great love and care. Trees were beginning to grow and there was green grass all around.... And the villagers had formed village level committees to take charge of the watershed development work.... The story of Jhabua shows that a beginning can be made and made quite fast.
>
> I call Jhabua outstanding because it is an effort to involve the people in land and water management on a scale and depth that no other government has attempted. Today, Jhabua is truly a temple of modern India, to use Nebruji's phrase; in fact, a temple of twenty-first century India, which shows how poverty can be eradicated from its roots by empowering the local people to manage their environment (Agarwal 1998).

What Agarwal talked about is the empowerment of the Bhil. As we have seen, the watershed project in Jhabua did that but the important question is: Can such a thing be replicated elsewhere?

For that to happen, we need to seek answers to two concerns that we had briefly alluded to. These are: (a) the dependency created by the civil society intermediaries, and (b) the relationship of the community institutions created by the watershed project to the Panchayati Raj institutions.

Dependency created by the civil society intermediaries

During interactions with the Bhil, it was clear that they were ambivalent about whether they would be in a position to manage things on their own without support from the civil society intermediaries. It is true that the civil society intermediaries did a great job in winning the trust of the Bhil. It was no mean task considering that the Bhil shy away from contact with the outside world and view outsiders with a great deal of suspicion and distrust. The civil society intermediaries managed to win their trust and confidence through sustained efforts. They gave the Bhil a sense of self-confidence and self-reliance, organised them into community constitutions, gave them technical assistance and built skills to make these institutions viable and effective. They helped these institutions to scale up their membership, range of functions, and awareness-raising. They created bridges between these Bhil institutions and the departments of the governments and outside agencies. All in all, the civil society groups fulfilled their strategic role in empowering the Bhil.

It was envisaged, however, that the civil society intermediaries would empower the Bhil during the project implementation period, but enable them to sustain the process after the project implementation was over. Why, then, do we have apprehensions about the Bhil sustaining the process after the civil service intermediaries withdraw? Why is that the project has not led to the creation of enduring institutions? In fact, this has been a weakness generally of the NGO-led programmes, in the sense that they have tended to foster a culture of dependence. It has been the common experience that processes and institutional arrangements have collapsed after the withdrawal of the NGOs (Shah 2006).

The fostering of such dependency is essentially a function of the duration of the project. At present, watershed projects are implemented over a four to five year programme, and this does not allow the time and space for newly formed local institutions to mature and take root. What is required for the purpose is to extend the duration of the watershed projects to eight years. The projects should have a phased approach, starting with a two year preparatory phase during which the basic resource survey could be completed, in addition to identification of problems and the required interventions. During the preparatory phase, the village community should be sensitised to the nature of the tasks involved, helped to set up the institutional mechanisms to handle those tasks and train those who will man the mechanisms.

The approval of the programmes for implementation which should be the second phase, should be subject to satisfactory completion of the preparatory segment. Whether the tasks of the preliminary phase have been satisfactorily completed need to be appraised by an external agency. During the implementation phase, which should be for four years, the action plans should be implemented, thereby creating appropriate conditions for the institutions to mature and take root. During this period, action plan for productivity enhancement and creating new opportunities for the third phase should be prepared. What is required is a system for continuous monitoring and appraisal of progress, impact assessment and research. These assessments should be the basis for decisions for approval and release of funds for graduating from one phase to the next and for different stages within each phase.

This was also the view of the watershed practitioners and was indicated in interactions with the NGOs in Jhabua. The watershed practitioners felt that the existing watershed guidelines of the government, which provided for a period of four years for watershed development including community capacity building and physical implementation, were too short. Given such a short timeline, the implementing agencies often concentrated on achieving physical targets, forgetting the far more intrinsically important role of community capacity building and participation. The watershed practitioners were of the view that the national watershed guidelines should provide for a longer period of watershed development, with two separate and distinct phases for capacity building and physical implementation. The second phase should follow only after a thorough evaluation of the first so as to ensure that the best returns come out of the investments made with the least associated risk.

This is also the view of the Technical Committee on watershed programme. In 2005, the Ministry of Rural Development, Government of India had constituted a Technical Committee under the chairmanship of S. Parthasarathy to review the performance of the watershed programmes in the country and suggest a possible way forward Shah (2006). The Parthasarathy Committee submitted its report, based on an extensive consultative process and a comprehensive review of available studies and reports on the subject. The Committee recommended that the duration of the watershed programmes should be extended to eight years.

With an extended duration, there is every reason to hope that there will be time and space for local institutions to mature and take root. It is important, however, to incorporate this aspect into the phase of awareness-raising and community capacity building: it needs to ensure that the villagers understand the criticality of taking charge of the institutions and managing them, without having to depend on continuing interventions from civil society intermediaries.

Relationship with Panchayati Raj institutions

Panchayati Raj Institutions (PRIs) are the officially designated agencies for rural development. In the Jhabua watershed project, as we have seen, the PRIs have been given important roles and responsibilities. The panchayat representatives are included in the Watershed Development Committees. The Gram Sabhas are consulted for the sanctioning of watershed works. However, interactions with the Bhil brought out that the ordinary Bhil had no confidence in the PRIs acting to promote their interest. Most Bhil were of the view that the Panchayats only promoted the private interests of their elite functionaries and not the collective interest of the community. Interactions with the Bhil, in fact, indicated an adverserial relationship between the PRIs and the watershed institutions.

As if to compound matters, the official thinking on the roles and responsibilities of the PRIs in watershed development programmes has not been particularly consistent. In 1994, the Hanumantha Rao Committee made a number of important recommendations to revamp the Drought Prone Areas Programme (DPAP) and Desert Development Programme (DDP) and suggested a set of common guidelines. The Committee suggested active encouragement of community participation and a greater role for the NGOs. This led to the Ministry of Agriculture making changes in the guidelines for National Watershed Development Project for Rainfed Areas (NWDPRA) to make it more participatory, equitable and sustainable. The PRIs were associated with the process only marginally, and, on the whole, it was an NGO-led process of watershed development (Vaidyanathan 2006).

The Hariyali guidelines of the Government of India came in 2003. The guidelines took a decisive stand on the roles and relationship between the Watershed Development Committees as separate entities, and the Gram Sabha and the Panchayats in planning and implementing decisions regarding watershed development programmes. In a sense, these guidelines were a wholesale reversal of the recommendations of the Hanumantha Rao Committee. In the Hariyali guidelines, the

Ministry of Rural Development abolished the separate and almost autonomous Watershed Development Committees and Watershed Associations, and handed over the watershed functions to the Gram Sabhas and the Gram Panchayats. In other words, the entire role and responsibilities of the watershed institutions were transferred to the PRIs (Vaidyanathan 2006).

What was the logic of the Hariyali guidelines? Ostensibly, it had something to do with considerations of equity and conflict resolution. Since one of the most important issues that arise in the case of watershed development programmes is that of conflict resolution and equity, the implementing agencies need to be equipped with appropriate legal and administrative authority to enforce their decisions (Shah 2006). The assumption was that the authorised role of PRIs could ensure this.

The Hariyali guidelines currently guide how the watershed development programmes in the country are implemented. How have they worked? The experience, which was reviewed by the Parthasarathy Committee, provided overwhelming evidence that the institutional arrangements sanctioned by the Hariyali guidelines have not worked well. The members of the panchayats are not in a position to discharge their responsibilities in respect of the watershed programme. Another weakness is that the secretaries of the panchayats are overloaded with so many diverse responsibilities of revenue, development and administration that it is unreasonable to expect them to find the time required of such quality and process intensity as demanded by the watershed programme (Shah 2006). As a result, watershed development has suffered a big setback.

What does the Parthasarathy Committee recommend? This Committee takes the middle ground of empowering the Panchayati Raj Institutions while at the same time getting the work done to meet the goals of the watershed programme. It also suggests restoring the key role of the Watershed Development Committees at the village level, but positioning them as one of the committees of the panchayat. In terms of the recommendations of the committee, the Watershed Development Committee should be elected in the meeting of the Gram Sabha and made to function as a committee of the panchayat (Shah 2006).

In other words, what the Parthasarathy Committee has done is empower the PRIs in the area of watershed development by widening their democratic base. Despite the huge problems of the panchayat-centred Hariyali guidelines, the Committee has not supported precluding the PRIs from watershed development. According to this Committee, the NGOs cannot hope to replace the government in watershed development, and it is hard to imagine the voluntary sector being able to upscale operations at the requisite level but even more importantly, because of questions of accountability in a democratic polity. But the Parthasarathy Committee sees a major role for civil society in the sense that doors are once again opened for NGOs who were virtually eliminated by the Hariyali guidelines (Shah 2006).

What would be the role of the NGOs in the scheme of things that the Parthasarathy Committee envisages? This Committee emphasises that the civil society should see its primary role in engagement with the state sector and panchayats rather than taking the lead role. It also thinks that the voluntary sector has a vital role to play in ensuring transparency and accountability of state institutions and in empowering the panchayats in close partnership with them. It is a pity that the Parthasarathy Committee should think so, because of the outstanding work that has been done by the NGOs in the area of watershed development, as we noted in the case of Jhabua. The NGOs have been oases of excellence and commitment in a vast desert of incompetence, corruption and lassitude.

The important question is: what would it mean for watershed development as an empowering strategy if the recommendations of the Parthasarathy Committee are accepted and clothed with the imprimatur of official policy? The answer is in the negative. If the panchayats are in charge of watershed development, empowerment of the poor through watershed institutions is not likely to take place for the simple reason that these institutions would reflect the character of the panchayats, which are, at present, captured by the elite and represent their interests. It would merely be a replay of how the Hariyali guidelines worked: the institutional arrangements were such a dismal failure.

Bibliography

Action for Social Advancement (1998), 'Two Years of Our Existence (July' 96–June '98)', Dahod.

Agarwal, Anil (1998a), 'The Story of Jhabua', *Down to Earth*, February 15, pp. 3–17.

Agarwal, Anil (1998b), 'Jhabua: Power from the People', *Down to Earth*, December 31, pp. 6–13.

Agarwal, Bina (1994), 'Gender and Legal Rights in Agricultural Land in India', *Economic and Political Weekly* 30: A 39–56.

BAIF (1987), *Land Management: Techniques of Soil and Water Conservation*, BAIF, Pune.

Bajpai, Nirupam and Jeffrey Sachs (1999), 'The Progress of Policy Reform and Variations in Performance at the Sub-National Level in India', Development Discussion Paper 730, Harvard Institute for International Development, Cambridge, Mass.

Bandura, A. (1981), 'Self-referent Thought: A Developmental Analysis of Selfefficacy', in J.H. Flavell and L. Ross (eds), *Social Logative Development: Frontiers and Possible Futures*, pp. 8–35, Cambridge, Cambridge University Press.

Bardhan, Pranab (1997), *The Role of Governance in Economic Development: A Political Economy Approach*, OECD Development Centre Study, Washington Centre Press, Washington, D.C.

Basu, Kaushik (1998), *Child Labour: Cause, Consequence, and Cure, with Remarks on International Labour Standards*, Policy Research Working Paper 2027, World Bank, Washington, D.C.

Bhatt, Ela (1989), 'Towards Empowerment', *World Development* 17(7): 1059–65.

Blair, Harry (2000), 'Participation and Accountability at the Periphery: Democratic Local Governance in Six Countries', *World Development* 28(1): 21–39.

Bonilla-Chacin, Maria and Jeffrey S. Hammer (1999), *Life and Death amongst the Poorest*, World Bank, Development Research Group, Washington, D.C.

Brown, David and Darcy Ashman (1996), 'Participation, Social Capital, and Intersectoral Problem Solving: African and Asian Cases', *World Development* 24(9): 1467–79.

Collector (2000), 'Human Development Pilot', Jhabua.

Collector, Jhabua (2005), 'Report on the Implementation of Watershed Project'.

Das Gupta, Monica (1995), 'Lifecourse Perspectives on Women's Autonomy and Health Outcomes', *American Anthropologist* 97(3): 481–91.

Desai, Mira (1996), *Social and Development Issues: Jhabua Development Communication Project,* Development and Educational Communication Unit (DECU), Indian Space Research Organisation, Ahmedabad.

District Census Office, *District Census Handbooks 1961, 1971, 1981 and 1991*, District Census Office, Jhabua.

District Rural Development Authority (1989), *Mukhya Bindu*, District Rural Development Authority, Jhabua.

Dodd, Pamela and Lorraine Guiterez (1990), 'Preparing Students for the Future: A Power Perspective on Community Practice', *Administration in Social Work* 14(2): 115–31.

Elwin, Verrier (1946), *Myths of Middle India*, Oxford University Press, Bombay.

Emerson, R.M. (1982), 'Power- Dependence Relations', *American Sociological Review*, 27, pp. 31–40.

French, J.R.P. and B. Raven (1959), 'The Bases of Social Power', in D. Cartwright (ed.), *Studies in Social Power*, pp. 79–91, Institute for Social Research, University of Michigan, Ann Arbor.

Government of India (1993), *Guidelines for Watershed Development*, Ministry of Rural Development, Government of India.

——— (2001), *National Human Development Report*, Planning Commission, Government of India.

——— (1997–98), *Health Information of India*, Central Bureau of Health Intelligence, Directorate General of Health Services, Ministry of Health and Family Welfare.

Government of Madhya Pradesh(1992), *Madhya Pradesh Rajya Ke Kshetriya Vikas Sanketak,* Madhya Pradesh Directorate of Economics and Statistics, Bhopal.

——— (1995), *The Madhya Pradesh Human Development Report (1995)*, Directorate of Institutional Finance, Govern-ment of Madhya Pradesh, Bhopal.

Government of Madhya Pradesh (1995), *Jhabua Jila Sankhyaki Pustika*, Bureau of Economics and Statistics, Government of Madhya Pradesh.

———— (1995), *Participatory Resource Mapping Approach*, Department of Rural Development, Bhopal.

———— (1995), *Madhya Pradesh Manav Vikash Prativedan*, Government of Madhya Pradesh, Bhopal.

———— (1998), 'Second Human Development Report', Directorate of Institutional Finance, Government of Madhya Pradesh, Bhopal.

———— (2002), 'Third Human Development Report', Directorate of Institutional Finance, Government of Madhya Pradesh, Bhopal.

Graham, D.C, *Sketch of the Bheel Tribes of the Province of Khandesh*, Part II, publisher unknown.

Harris-White, Barbara (1996), *The Political Economy of Disability and Development with Special Reference to India*, United Nations Research Institute for Social Development Discussion Paper 73, Geneva.

Hendley, Major (1908), *The Account of the Bhils,* Self-published.

Impact Evaluation of Watershed Development Activities in Jhabua District of Madhya Pradesh Using Remote Sensing Techniques (1998), Remote Sensing Applications Centre, M P: Council of Science & Technology, Bhopal.

Institute of Resource Conservation (1995), *Training Manual for Watershed Secretary*. Jhabua: Institute of Resource Conservation.

ISRO (1995), *Jhabua District of Madhya Pradesh at a Glance*. Ahmedabad: Development and Educational Communication Unit (DECU), Indian Space Research Organisation.

Iyer, Pico (1993), *Falling Off the Map*, Viking, Penguin India, New Delhi.

Jejeebhoy, Shireen (1995), *Women's Education, Autonomy, and Reproductive Behaviour: Experience from Developing Countries*, Oxford University Press, New York.

Joy, K.J, Amita Shah, Suhas Paranjape, Shrinivas Badiger and Sharachchandra Lele (2006), 'Issues in Restructuring', *Economic and Political Weekly,* July 8–15, pp. 2994–96.

Joshi, Deep (2006), 'Broadening the Scope of Watershed Development,' *Economic and Political Weekly,* July 8–15, pp. 2987–91.

Khandker, Shahidur (1998), *Fighting Poverty with Microcredit: Experience in Bangladesh*, Oxford University Press, New York.

Kochar, Anjini (1999), 'Smoothing Consumption by Smoothing Income: Hours-of-Work Response to Idiosyncratic Agricultural Shocks in Rural India', *Review of Economics and Statistics,* 81(1): 50–61.

Krishna Kumar, K.N. (1997), *Forestry: Jhabua Development Communication Project*, Indian Institute of Forest Management, Bhopal.

Lips, H. (1991), *Women, Men and Power*, Mayfeld, Mountain view, CA.

Literary Mission (1995), *Patey Ki Baat-A Play on Soil and Water Conservation*, District Collectorate, Jhabua.

Lvard, C.E. (1909), 'The Jungle Tribes of Malwa', Monograph No. 11, The Ethnological Survey of Central Indian Agency.

Mahajan, Vinaya (1996), *Needs Analysis and Communications Brief Writing for Watershed Management: A Situation Study*, Management, Research and Planning Group, Ahmedabad.

Malcolm, John (1880), *A Memoir of Central India Including Malwa and Adjoining Provinces*, vol I, Parbury and Allen, London, Kingsburg.

Manor, James (1999), *The Political Economy of Democratic Decentralization*. Directions in Development Series, The World Bank, Washington, D.C.

Moser, Caroline (1998), 'The Asset Vulnerability Framework: Reassessing Urban Poverty Reduction Strategies', *World Development*, 26(1): 1–19.

NABARD (2001), 'Madhya Pradesh State Focus Paper', National Bank for Agriculture and Rural Development, Bhopal.

NCHSE (1994), *PRA Report*, National Centre for Human Settlements and Environment, Village Gopalpura.

——— (1992), *Role in Human Settlements and Meeting the Environmental Challenges*, National Centre for Human Settlements and Environment, Bhopal.

——— (1993), *Sustainable Utilisation of Natural Resources in Jhabua District*, National Centre for Human Settlements and Environment, Bhopal.

——— (1996), *Vacham, Watershed Management Issue* (January), National Centre for Human Settlements and Environment, Bhopal.

Naik, T.B. (1956), *The Bhil- A Study*, Admijati Sewak Sangh, Delhi.

Narayan, Deepa (2002), *Empowerment and Poverty Reduction: A Source Book*, The World Bank, Washington D.C.

Narayan, Deepa, Robert Chambers, Meera K. Shah and Patti Petesch (2000), *Voices of the Poor: Crying Out for Change*, Oxford University Press, New York.

Narayan, Deepa, with Raj Patel, Kai Schafft, Anne Rademacher and Sarah Koch-Schulte (2000), *Voices of the Poor: Can Anyone Hear Us?* Oxford University Press, New York.

NIRD (1987), 'Background Papers', International Seminar on Agricultural Communication, Hyderabad.

National Remote Sensing Agency (NRSA). 2002. 'Integrated Mission for Sustainable Development', Department of Space, Government of India.

Parenti, M. (1978), *Power and the Powerless*, St. Martin's Press.

Page, Nanette and Cheryl Czuba (1999), 'Empowerment: What is it?' *Journal of Extension Vol 37 Number 5*.

Participatory Irrigation Management Conference (1996), 'Report', Water and Land Management Institute, Bhopal.

PRAYAS (1999), *A Modest Beginning: A Documentation of the Programmatic Interventions of PRAYAS in 1998–1999*, Organisation for Sustainable Development, Dahod.

Rajiv Gandhi Watershed Mission (1999), *Intermediate Assessment of Watershed Management Programmes in Jhabua District. Jhabua:* Rajiv Gandhi Watershed Management Mission.

Rapid Household Survey (1999), Ministry of Health and Family Welfare and Centre for Operations Research and Training, Vadodara.

RGMWD (1995), *Geet-Naatak: Street Plays on Watershed Management*, Rajiv Gandhi Mission for Watershed Development, District Jhabua.

Roy, S.C. (1970), *The Mundas and their Country*, Asia Publishing House Bombay.

Russel, R.V. (1908), *Central Provinces District Gazetters, Nimhar District, Volume A,* Pioneer Press, Allahabad.

Russel, R.V., and Hiralal (1916), *Bhil, The Tribes and Castes of the Central Provinces of India*, Macmillan, London.

Sah, Raaj K. (1991), 'Fallibility in Human Organizations and Political Systems', *Journal of Economic Perspectives* 5 (2): 67–88.

Sahay, Sushama (1998), *Women and Empowerment: Approaches and Strategies*, Discovery Publishing House, New Delhi.

Sen, Amartya (1985), 'Well-being, Agency and Freedom: The Dewey Lectures 1984', *Journal of Philosophy,* 82: 181–90.

——— (1992), *Inequality Reexamined*, Clarendon, Oxford.

SEWA (1997), *Liberalizing for the Poor,* Self Employed Women's Association, Ahmedabad.

Shah, M. (2006), 'Towards Reforms', *Economic and Political Weekly*, July 8–15, pp. 2981–84.

Sharma, Amita (2000), 'Idea of Education', Occasional Paper No. 7, Rajiv Gandhi Mission, Bhopal.

Sherring, M.A. 1872 (1974 rpt), *Hindu Tribes and Castes Together with an Account of the Mohammedan Tribes of the Central Provinces*, Delhi.

Shrivastava, Manoj (1999), *Bairani Kuldi: Women Thrift and Credit Groups of Jhabua*, Collectorate, Jhabua.

Shrivastava, Manoj and S.K. Shukla (1997), *Gramkosh*, Collectorate, Jhabua.

Singh A. K. , P. K. Sahoo, U. C. Vashishtha, P. Singh and Sachin Narshana (1996), *Needs Analysis and Communication Brief Writing for Education: A Situation Study*, Audio Visual Research Centre, Devi Ahilya Vishwavidyalaya, Indore.

Sisodia, Y.S. (1996), *Panchayati Raj: Jhabua Development Communications Project*, Institute of Social Science Research, M P. Ujjain.

The World Bank (2000), 'World Development Report', The World Bank, Washington, D.C.

Todd, Colonel James (1920), *Annals and Antiquities of Rajasthan*, vol. 1, (ed.) William Crooke, Oxford University Press, London.

Uberoi, Patricia (1999), *Gender and State Policies in India*, 1955–2000, The World Bank, Washington, D.C.

UNDP (1999), 'Human Development Report', Oxford Univeristy Press, New York.

UNESCAP (2006), 'The Empowerment of the Rural Poor through Decentralization in Poverty Alleviation Actions', http://www.unescap.org/

Uphoff, Norman (1993), 'Grassroots Organizations and NGOs in Rural Development: Opportunities with Diminishing States and Expanding Markets', *World Development*, 21(4): 607–22.

Uppal, D.K. (1996), *Needs Analysis and Communication Brief Writing for Agriculture: A Situation Study of Jhabua District of Madhya Pradesh*, Makhanlal Chaturvedi Rashtriya Patrakarita Vishwavidyalaya, Bhopal.

Vaidyanathan, A. (2006), 'Restructuring Watershed Development Programme', *Economic and Political Weekly*, July 8–15, pp.2984–87.

Varma, S.C. (1978), *The Bhil Kills,* Self-published.

Weber, M. (1946), *From Max Weber,* H. H. Gerth and C. W. Mills (eds), Oxford University Press, New York.

Index

Printed in the United States
by Baker & Taylor Publisher Services